PROBLEMS IN INORGANIC CHEMISTRY

By

Dr. Syed Aftab Iqbal

M.Sc. , Ph.D., FICS

FICC, FIAEM, MNASc.

Professor

Department of Chemistry

Saifia Science College

Barkatullah University

Bhopal (India)

DISCOVERY PUBLISHING HOUSE PVT. LTD.

NEW DELHI-110 002

Published by:
Tilak Wasan
DISCOVERY PUBLISHING HOUSE PVT. LTD.
4831/24, Ansari Road, Prahlad Street
Darya Ganj, New Delhi-110002 (India)
Phone: +91-11-23279245, 43764432
Fax: +91-11-23253475
E-mail: parul.wasan@gmail.com
info@discoverypublishinggroup.com
discoverypublishinghouse@gmail.com
web: www.discoverypublishinggroup.com

***First Edition:* 2011**
ISBN: 978-81-8356-817-3

Problems in Inorganic Chemistry

Printed at:
Mehra Offset Press
Delhi

Preface

The book "**Problems in Inorganic Chemistry**" has been written the requirements of graduate students of all Indian Universities. It covers the complete syllabus of physics prescribed by Technical Universities. It covers the complete syllabus of electronics prescribed by Technical Universities. I have spared no pains in applying my long experience of degree class teaching to make the book useful to the students.

The subject matter of this book has been presented in this book has sufficient theory and maximum solved examples. One of salient course syllabi features of the present edition of the book is the inclusion of a large number of typical worked out problems which will elucidate various abstract principles and theories discussed in the text.

The author in general hope. That the present book will be warmly received by the students and teachers. We shall indeed be very thankful to our colleague for their recommendations this book of for their students.

—*Author*

CONTENTS

4. Models and Properties of Nuclear

Introduction, Classification of Nuclear, Nuclear Size, Electrical and Magnetic Properties of the Nucleus, Electric Quadropole Moment, Bose-Einstein Statistics, Atomic Mass Unit, Mass Defect, Significance of Packing Fraction, Nuclear Isomerism, Binding Energy, Isotopes, Separation of Isotopes, Composition of Nucleus, Elementary Ideas About Nuclear Forces, Models of the Nucleus, Warner's Theory of Coordination, Experimental Evidence in Support of Werner's Theory, Crystal Field Splitting by Different Geometrically, Arrangements of Ligands, Applications of Crystal Field Theory, Octahedral Complexes, Co-ordination Compounds, Some Solved Problems.

1

Basic Problems of the Atomic Structure

DERIVATION OF THE RUTHERFORD'S FORMULA

The experiments on the scattering of α-particles confirmed Rutherford's theory so well that the nuclear atom model was at once universally accepted but Rutherford himself soon realised that his model was not free from limitations, the chief among them arising from considerations of the distribution of the electrons outside the nucleus and the stability of the atom as a whole. For, it became obvious that in the nuclear atom, *equilibrium could not be secured by the operation of electrostatic forces* alone between the positively charged nucleus and the negatively charged electrons outside the nucleus. For instance, considering the case of an atom with two electrons, the nuclear charge is +2e. If the electrons are symmetrically placed at a distance r from the nucleus, the force of attraction between the nucleus and each of the electrons is $\frac{2e^2}{4\pi \epsilon_0 r^2}$ while the force of repulsion between the electrons is $\frac{e^2}{4\pi \epsilon_0 (4r^2)}$ Since the force of attraction is eight times greater than that of repulsion, the condition of stability is not satisfied and the electrons will fall into the nucleus, thus destroying the stable structure of the atom. In order to overcome this difficulty, Rutherford suggested that *electrons might be assumed to revolve round the nucleus*, like the planets round the sun, at such a speed that the mechanical centrifugal force would just balance the net excess of electrostatic attraction and in consequence stability of the atom could be secured.

But such an assumption brought in its wake a very serious difficulty from the point of view of the electromagnetic theory, according to which *a revolving electron should radiate energy continuously.* Now, this energy can only come from the atomic system, which will therefore steadily lose energy. As a result, the electron will approach the nucleus by a spiral path giving out radiation of constantly increasing frequency and finally fall into the nucleus. Thus, the orbital motion of the electron destroys the very purpose for which it was postulated *viz.*, the stability of the atom. Further, emission of radiation of constancy increasing frequency has no experimental support whatsoever, since elements are actually found to emit discrete spectral lines of definite frequencies. One is therefore forced to conclude that the Rutherford nuclear atom model with revolving electrons is defective or the classical electromagnetic theory fails in the present case. The dilemma was solved in 1913 by Niels Bohr, who admitting the failure of the classical theory applied with remarkable success the quantum theory to the Rutherford nuclear atom model with revolving electrons.

An α-particle moving along PO approach the relatively heavy nucleus, stationary at N. Since both are positively charged, there is a force of repulsion between them, which being governed by the Coulomb's law of inverse squares, will increase enormously as the α-particle gets closer to the nucleus. Applying the properties of motion in central orbits to the α-particle thus repelled by the nucleus, it can be shown that the path of the α-particle will, in general, change from a straight line to a hyperbola PAP', one of whose foci is N and whose asymptotes PO give the initial and final directions of the α-particle. In deriving the scattering formula the following assumption are made:

(i) The wave aspect of the atom being neglected and considered that call is an between two particles, α-particles and nucleus.

(ii) The nucleus is considered to be so heavy that its impact may be disregarded.

(iii) The nucleus and the α-particle are considered as point charges, *i.e.*, mere centres of Coulombian force; thus, the dimensions of the interacting particles are not taken into account.

Let the perpendicular distance of PO from N be p, *i.e.*, the shortest distance from the nucleus to the initial direction, which is called the *impact parameter*. Let m be mass of the α-particle, 2e its charge and υ_0 its initial velocity. Let Z be the atomic number of the element that scatters the α-particles, so that Ze is the charge on the nucleus. The angle of deviation or scattering of the α-particle is evidently

The magnitude of ϕ will depend on several factors, such as the charge of the nucleus Ze, the charge of the α-particle 2e, its mass m, its velocity υ_0 and impact parameter P.

To derive an expression for ϕ, let us consider the case of a central impact, *i.e.*, when the α-particle is directed straight towards N so that p = 0. On account of the repulsive force, the α-particle will be stopped at a certain distance b from the nucleus N and made to retrace its path, in which case ϕ is equal to 180°. This distance of closest approach b can be determined by using the principle of conservation of energy.

The electrostatic potential at a distance b due to the nucleus

$$= \frac{Ze}{4\pi \epsilon_0 b}$$

This acts on a charge 2e of the α-particle. Hence the potential energy of an α-particle when it is at the distance b from the nucleus

$$= \left(\frac{Ze}{4\pi \epsilon_0 b}\right)2e = 2\frac{Ze^2}{4\pi \epsilon_0 b}$$

Since the α-particle is momentarily stopped at the distance b, its initial kinetic energy is completely changed into potential energy. Neglecting the loss of energy due to interaction with the peripheral electrons

$$\frac{1}{2}m\upsilon_0^2 = \frac{2Ze^2}{4\pi \epsilon_0 b}$$

or $$b = \frac{Ze^2}{\pi \epsilon_0 m\upsilon_0^2} \qquad ...(1)$$

As it is not possible, in practice, to direct the α-particle exactly towards the nucleus, we must consider the case when p ≠ 0. In such a case, the α-particle will be deflected through an angle ϕ, which is less than 180° and will travel along the hyperbolic path PAP'

Let V be the velocity of the α-particle at the vertex A. Using the principles of conservation of energy and of momentum and taking distance NA = d

$$\frac{1}{2}m\upsilon_0^2 = \frac{1}{2}m\,V^2 + \frac{2Ze^2}{4\pi \epsilon_0 d} \qquad ...(2)$$

$$m\upsilon_0 p = mVd \qquad ...(3)$$

Substituting from (1), the value of $2Ze^2$, *viz.*, $2\pi\in_0 bm\upsilon^2_0$ in equation (2), we get

$$\frac{1}{2}m\upsilon_0^2 = \frac{1}{2}m\ V^2 + \frac{2\pi\in_0 bm\upsilon_0^2}{4\pi\in_0 d}$$

$$\therefore \qquad V^2\ \upsilon_0^2\left(1-\frac{b}{d}\right) \text{ or } \frac{V^2}{\upsilon_0^2}\left(1-\frac{b}{d}\right)$$

From equation (3), we get

$$p^2 = \frac{V^2}{\upsilon_0^2}d^2 = d^2\left(1-\frac{b}{d}\right) = d(d-b)$$

Using the properties of the hyperbola, *viz.*,

$$\in = \frac{1}{\cos\theta}, \text{ where } \theta = \left(\frac{r-\phi}{2}\right), \text{ and On } = \in.\text{OA}.$$

$$\text{or} \qquad d = NO + OA = NO\left(\frac{1+1}{\in}\right) = NO(1+\cos\theta)$$

$$= \left(\frac{p}{\sin\theta}\right)(1+\cos\theta) = p\left\{\frac{(1+2\cos^2(\theta/2)-1}{(2\sin\theta/2\cos(\theta/2)}\right\}$$

$$= \frac{p.\cot\theta}{2}$$

$$\therefore \qquad p^2 = p\left(\frac{\cot\theta}{2}\right)\left[p\ \cot\left(\frac{\theta}{2}\right)-b\right]$$

$$\text{or} \qquad p = p\left[\cot^2\left(\frac{\theta}{2}\right)\right]-b\cot\frac{\theta}{2}$$

$$\therefore \qquad b = p\frac{[\cot^2\theta/2)-1]}{\cot\theta/2} = 2p\ \cot\theta = 2p\cot\left(\frac{\pi-\theta}{2}\right)$$

$$\text{or} \qquad b = \frac{2p\tan\phi}{2}$$

Substituting for b from equation (1)

$$\tan\frac{\phi}{2} = \frac{b}{2p}\frac{2Ze^2}{4\pi\in_0 m\upsilon_0^2 p} \qquad ...(4)$$

This relation shows that Z, m and υ_0 being kept constant, as the impact parameter p decreases from relatively high value upto the limit zero, ϕ increases from 0 to 180°. This means that when the α-particle passes for away from the nucleus (*i.e.*, p is large) the angle of scattering is very small. As the distance p diminishes, *i.e.*, as the α-particle passes closer and closer to the nucleus, the angle of scattering will become larger and larger and in the limiting case of central impact (p = 0), ϕ = 180°, *i.e.*, the α-particle will be forced to retrace its path after approaching the nucleus upto a distance b.

We shall now deal with the case realised in the actual experi-ment where a narrow beam of α-particles is incident normally on a thin foil of the scatterer, and the scattered particles are detected by means of scintillations produced by them on a fluorescent screen normal to the direction of view. Let n be the number of atoms per unit volume of the scatterer of thickness t. Let Q be the total number of a-particles that strike unit area of the scatterer. From simple probability considerations the number of α-particles N that are scattered through an angle ϕ and strike unit area of the fluorescent screen S at a distance r can be estimated as follows:

The scattered atoms is distributed, the probable number of α-particle within the distance of the impact parameter p of a nucleus is given by

$$\pi p^2 Qtn \qquad ...(5)$$

Hence the number of α-particles having an impact parameter between p and p + dp is

$$d(\pi p^2 Qtn) = 2\pi Qntpdp$$

After impact, these particles will be deflected through an angle between ϕ and $\phi + d\phi$. Thus the number of α-particles scattered between angles ϕ and $\phi + d\phi$ is

$$2\pi Qntpdp \qquad ...(6)$$

From equation (4),

$$p = \left(\frac{b}{2}\right)\cot\frac{\phi}{2} \text{ and } dp = \left(\frac{b}{2}\right)\left(-\frac{1}{2}\text{cosec}^2\frac{\phi}{2}\right)d\phi$$

$$= -2\,\pi Q n t \times \left(\frac{b}{2}\right)\cot\frac{\phi}{2} \times \left(\frac{b}{4}\right)\text{cosec}^2\left(\frac{\phi}{2}\right)d\phi$$

sharp bright coloured lines. These lines are the images of the slit of the spectroscope formed by lights of different colours. The entire series of images is called a 'line spectrum'. The different lines differ in intensity and nature. Some are sharp, some are sharp on one side and diffused on the other, and others are diffused on both sides.

The line spectrum is the characteristic of the atom or the ion. It means that a particular atom or ion always gives the characteristic set of spectral lines, and no two atoms or ions can give the same spectral line. For example, sodium atom gives two intense yellow lines called D_1 and D_2 lines.

ABSORPTION SPECTRA

The absorption spectra are obtained when the absorbing substance is placed between a source emitting a continuous spectrum and the slit of the spectroscope. In such cases, certain colours (wavelengths) are absorbed by the substance. Hence the spectrum is found to consist of dark lines or bands against a bright background. An example of such spectra is sun's spectrum. It is a line absorption spectrum. It was studied in detail by Fraunhoffer offer who named the absorption lines as A, B, C, D... These lines inform us about the elements which are present around the sun.

Similarly, when an intense beam of continuous white light is passed through sodium vapour and then sent into a spectroscope, we obtain two dark lines on a continuous background in the same positions as the yellow D_1 and D_2 lines in the sodium, emission spectrum. If the sodium vapour is replaced by Iodine vapour (I2), an absorption band spectrum of I2 molecule is obtained.

The Sommerfeld Model

Inspite of many success, Bohr's theory was found to be inadequate to explain certain details in the spectrum of hydrogen. For instance, the Hx line of the Balmer series was found to contain several components. This fine structure of spectral lines could not be explained by Bohr's theory which assumed that there was only one orbit for each quantum number, whereas the observed fine structure suggested that for any given quantum number n there might be several orbits of slightly different energies.

Sommerfeld, in 1915, guided by the above suggestion modified Bohr's theory by introducing the following modifications.

(a) Concept of elliptical orbits, and

(b) Relativistic variation of the mass of electron.

We will discuss these modifications one by one.

(a) *Elliptical Orbits :* In order to explain the multiplicity of spectral lines. Sommerfeld introduced the concept of elliptical orbits. He postulated that :

(i) Since the electron is moving around and under the influence of a massive nucleus, like a planet around the central massive sun, it might describe elliptical orbits as well.

(ii) While retaining the first circular orbit suggested by Bohr, Sommerfeld assumed one additional elliptical orbit in the case of Bohr's second orbit and added two additional elliptical orbits to Bohr's third orbit and so on Fig. 1.

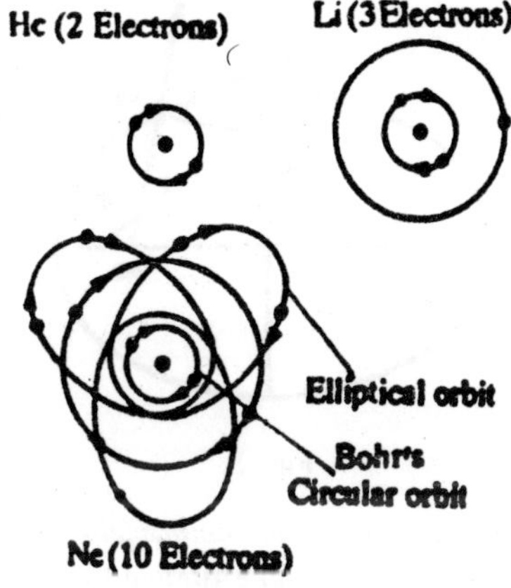

Fig. 1

(iii) The nucleus is one of the foci for all these orbits.

(iv) In an elliptical orbit, the major and minoraxes will differ in lengths, but as the orbit broadens, they will approach each other and becomes equal when the orbit becomes circular. Circular orbit is only a special case of the elliptical orbit. The orbit is defined by two quantum numbers, *i.e.*, n and k which correspond to the major and minor axes of the ellipse. They are related as follows.

$$\frac{\text{Principal quantum number}}{\text{Azimuthal quantum number}} = \frac{n}{k} = \frac{\text{length of major axis}}{\text{length of minor axis}}.$$

It is clear from the above that for any given value of n, k cannot be zero as in that case the ellipse would degenerate into a straight line passing through the nucleus. Further, k cannot be more than n since b is always less than a. When n = k the path becomes circular. This is a limit to the number of different orbits that an electron may have in any energy level.

Such possible paths are referred to as sub-levels. For a given value of n, k can have only n different values which mean that there may be only n elliptical orbits or sub levels with different eccentricity. For elliptical orbits the values of k would be (n – 1), (n – 2) etc., down to k = 1. When k = 0, the ellipse would be a straight line. When n = 3, k = 3, (circular), 2, 1, two of sommerfeld model lines in its subdivision of the original Bohr stationary levels into various sub-levels of slightly differing energies as given by difference in orbit shapes.

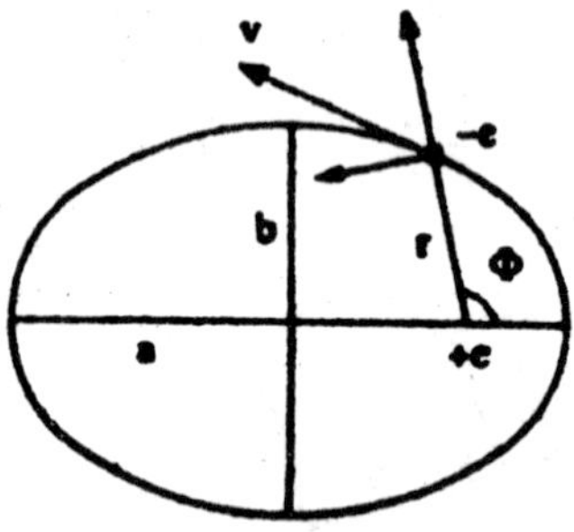

Fig. 2

Simple mathematical approach to Sommerfeld model : Let us consider the motion of an electron (– c) in an elliptical orbit as shown in Fig. 2. Its position at any instant can be fixed in terms of polar coordinates, r and f where r is the distance of the electron from the nucleus (+ e) at one of the foci of the ellipse and f is the angle which the radius vector makes with the major axis of the ellipse. The tangential velocity v of the moving electron at any instant can be resolved into two components.

(i) One radial, *i.e., dr/dt* along the radius vector. Corresponding to this there will be a radial momentum Pr equal to m (dr/dt).

(ii) Other transverse, *i.e.*, m at right angles to the radius vector equal to r (df/dt). Corresponding to this there is angular or azimuthal momentum of equal to mr^2 (df/dt), where m is the mass of the electron.

As Sommerfeld considered the circular orbits to be special cases of elliptical orbits, he assumed that the elliptical orbits should satisfy the quantum condition of Bohr just as the circular orbits, *i.e.*,

$$\oint p_r\, dr = n_r\, h \qquad ...(1)$$

$$\oint p_f\, d\phi = n_\phi\, l_t \qquad ...(2)$$

Thus, the single n of Bohr's theory has been replaced by the two new quantum numbers n_r and n_ϕ, *i.e.*,

$$n = n_r + n_\phi \qquad ...(3)$$

where n_r is the radial quantum number and nf is the angular or azimuthal quantum number. The total energy E is given by

$$E = \text{P.E.} + \text{K.E.}$$

$$= \text{P.E.} + \text{Radial K.E.} + \text{Angular K.E.}$$

$$= -\frac{e^2}{r} + \frac{1}{2} m \left(\frac{dr}{dt}\right)^2 + \frac{1}{2} mr^2 \left(\frac{d\phi}{dt}\right)^2 \qquad ...(4)$$

From equations (1), (2), (3) and (4), it can be shown that

$$1 - e^2 = \frac{b^2}{a^2} = \frac{n_\phi^{\ 2}}{(n_\phi + n_r)^2}$$

where e is the eccentricity of the ellipse whose semi-major and semi-minor axes are a and b respectively.

and $$E = -\frac{2\pi^2 mZ^2 e^4}{h^2}\left(\frac{1}{n_\phi + n_r)}\right)^2 \qquad ...(6)$$

From equation (5), we get

$$(1 - e^2)^{1/2n} = \frac{b}{a} = \frac{n_\phi}{n_\phi + n_r} = \frac{n_\phi}{n}.$$

Thus, equation (6) can be written as

$$E = \frac{2\pi^2 mZ^2 e^4}{n^2 h^2} \qquad ...(7)$$

From equation (7), it follows that

(i) The elliptical orbits which have the same value of n, though of different eccentricities, have the same energy. It is not correct.

(ii) All orbits having the same values of the semi major axis possess the same energy, since the length for the semi-major axis is determined solely by the total quantum number n ($= n_f + n_r$). Hence the energy of any of these permitted elliptical orbits is identical with that of a circular Bohr orbit, whose radius is equal to the semi-major axis of the ellipse.

From the above it follows that the theory of elliptical orbits in spite of two new quantising conditions involved, introduces not new energy levels other than those given by Bohr's theory of circular orbits. No new spectral lines, which would explain the fine structure, are there fore predicted. Thus, Sommerfeld has to modify his own model by suggesting the variation of mass of electron with velocity called relativistic variation.

Relativistic Variation of the Mass of Electron : The velocity if an electron moving in an es path of the electron is found to be no longer a simple ellipse, it is indeed, no longer a closed figure, but is transformed into a complicated curve known as rosette –a processing ellipse Fig. 3. The total energy E of the system corrected by the relativistic variation of the mass of electron can be shown to be

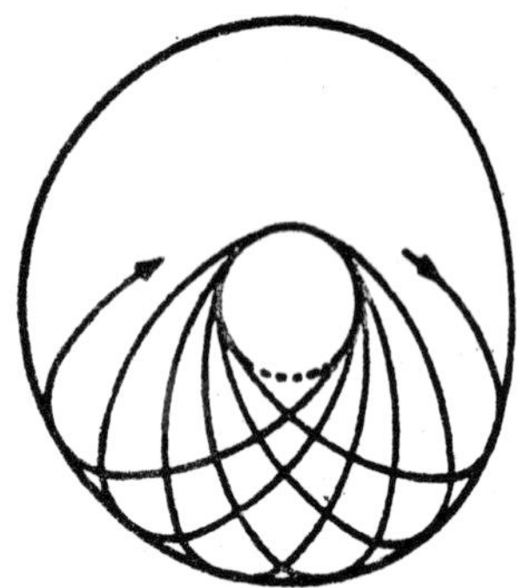

Fig. 3 : Sommerfeld elliptical orbit precessing about an axis through one of its focci.

$$E = -\frac{2\pi^2 mZ^2e^4}{h^2}\left[\frac{1}{n^2} + \frac{4\pi^2e^4Z^2}{c^2h^2}\left(\frac{n}{n\phi} - \frac{3}{4}\right)\frac{1}{n^4} + \dots\right] \quad \dots(8)$$

The relativistic correction, therefore, results in splitting up a given energy level E_α into n levels differing slightly from one another in energy. The splitting up of each energy levels gives rise to a fine structure of single spectral line, one application of the usual Bohr frequency conditions, When one explains the fine structure of lines, one should not take into account all theoretical possible transitions of the electron from one elliptical orbit to another that actually occur. According to a principal known as the selection rule, transition can take place only between orbits for which the azimuthal quantum number changes by + 1 or – 1, *i.e.*, nf = ± 1.

Limitations of Sommerfeld model : (i) Though Sommerfeld's theory is fairly well verified yet it leads only to three components for the structure of Hx line, while there should be really five. Thus the relativistic atom model has met with a partial success.

(i) The concept of elliptical orbits due to Sommerfeld gives the correct total (n) of possible azimuthal quantum number, but the actual values are not correct. The experimental studies as well as theoretical treatment based on wave mechanics show that azimuthal quantum number can be zero, so that the values can be 0, 1, 2,... –1, thus making a total of n possibilities. The correct and new azimuthal quantum number is denoted by 'l' to avoid the confusion. Thus, l is equal to K – 1.

(ii) It provides no idea about the number of electron which can be accommodated in a particular orbit.

(iii) It provides macerate values for angular momentum.

(iv) Sommerfeld model could not explain Zeeman and Stark effect.

Explanation of fine details in line spectrum by Sommerfeld model : It can be explained as follows.

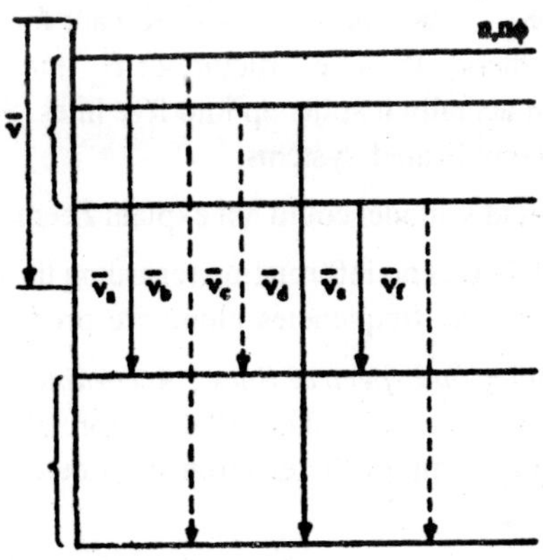

Fig. 4

When an electron moves in an elliptical path there will be a displacement each time in its elliptical path and thus resulting in a small difference in energy. Thus, the path of the moving electron is no longer a simple closed ellipse but a complicated curve, known as rosette, which is made up of different elliptical orbits of slightly different energies. This *difference in energies of elliptical orbits explains the existence of components of spectral lines in the spectrum of higher elements.*

Criticism of the theory : (1) Suppose there are two orbits with principal quantum numbers, n = 2 and n = 3, Fig. 4 respectively. Consider an electron which falls from third to the second orbit. There may be six possible transitions, each giving one fine line. But actual observations yield only five lines.

Possibilities.

(i) $3K_3 \rightarrow 2K_2$ *i.e.*, $\Delta K = 3 - 2 = +1$

(ii) $3K_2 \rightarrow 2K_1$ *i.e.*, $\Delta K = 3 - 1 = +2$

(iii) $3K_2 \rightarrow 2K_3$ *i.e.*, $\Delta K = 2 - 2 = 0$

(iv) $3K_2 \rightarrow 2K_1$ *i.e.*, $\Delta K = 2 - 1 = +1$

(v) $3K_1 \rightarrow 3K_2$ *i.e.*, $\Delta K = 1 - 2 = -1$

(vi) $3K_1 \rightarrow 2K_1$ *i.e.*, $\Delta K = 1 - 1 = 0$

where ΔK indicates the change in azimuthal quantum number. Sommerfeld applied *selection rule* which limited the number of transitions. According to such rule only those transitions are possible for which the quantum number K changes by – 1 or + 1 *i.e.*, $\Delta K = \pm 1$. Hence transitions like ($3K_1 \rightarrow 2K_1$), ($3K_3 \rightarrow 2K_1$) and ($3K_2 \rightarrow 2K_2$) are forbidden. Therefore, according to this theory, the fine structure of H_α lines should made only of three lines. But actually it splits up into five lines. Sommerfeld theory fails to explain complicated systems.

(2) Sommerfeld's model could not explain Zeeman and Stark effect.

(3) The model gives no information regarding the relative intensities of lines, whose frequencies alone are predicted.

Wave mechanics and spectral lines : According to wave mechanics orbits do not exist in the atom. Thus, the interpretation of the emission of radiation due to a jump of the electron from outer to the inner orbits does not hold good.

In wave mechanics, place of orbits has been taken by stationary states of atoms with definite energies. When any state excited by the absorption of energy comes to the ground state, spectral lines are produced. Thus, *the frequency of each spectral line may be regarded as a 'beat' frequency between two states of the atom, which gives the same result as that of Bohr.*

As Ψ^2 represents the electrical charge density, then in the stationary state this remains constant and so also the charge density. But when any state becomes excited by the absorption of energy, Ψ is no longer constant and varies periodically and so also the charge density. *This periodic variation in charge density is accompanied by the emission of radiation.*

Excitation and Ionisation Potentials of an Atom : According to the Bohr's theory, in an atom there are certain discrete orbits only in which an electron can revolve without radiating energy. Each of these orbits of a given atom is characterised by a certain (quantised) energy. Hence we say that there are certain discrete energy levels in an atom.

When an electron in an atom absorbs sufficient energy from an outside source, it rises from its present energy level to an higher level. The atom is than said to be 'excited'. This excited state lasts only for about 10^{-8} second after which the electron jumps back to the inner level, emitting the absorbed energy is emitted as electromagnetic radiations.

The energy required to excite or to ionise a given atom is perfectly definite.

For example, in case of hydrogen atom, the energy of the nth orbit is

$$E_n = -\frac{me^4}{8\varepsilon_0^2 h^2}\left(\frac{1}{n^2}\right).$$

Substituting the known values of

$$m\ (= 9.1 \times 10^{-31}\ \text{kg}),$$

$$e\ (= 1.6 \times 10^{-19}\text{C}),$$

$$h\ (= 6.62 \times 10^{-34}\ \text{J-s})$$

and ε_0 ($= 8.85 \times 10^{-12}\text{C}^2/\text{N m}^2$), we get

$$E_n = -2.17 \times 10^{-18}\left(\frac{1}{n^2}\right)\ \text{joule (J)}$$

$$= -\frac{2.17 \times 10^{-18}}{1.6 \times 10^{-19}}\left(\frac{1}{n^2}\right)$$

$$= -\frac{13.6}{n^2} \text{ eV.} \qquad [1 \text{ eV} = 1.6 \times 10^{-19} \text{ J}]$$

Putting n = 1, 2, 3, ... ∞, the energies of the 1st, 2nd, 3rd, ... ∞ energy levels come out to be – 13.6, – 3.4 – 1.51, ... 0 electron volts respectively.

Hence the energy to be supplied to the atom to raise an electron from the first to the second orbit is (13.6 – 3.4) = 10.2 electron-volts, from first to the third orbit is (13.6 – 1.51) = 12.09 electron volts,* and to the ionized state is

HYDROGEN SPECTRUM PRINCIPLE OF BOHR'S THEORY

The existence of sharp spectral lines cannot be explained from the classical electro-magnetic theory. Bohr explained it by applying Planck's quantum hypothesis to the Rutherford's atomic model.

The Rutherfod's atom consists of a central massive nucleus containing the positive charge of the atom, and the electrons move round the nucleus in circular planetary orbits. The centripetal force required for the orbital motion is provided by the electrostatic attraction between the positively-charge nucleus and the negatively-charged electron.

Bohr proposed two postulates.

(i) An electron can move only in those orbits for which the angular momentum L of the electron is an integral multiple of h/2π where h is Planck's constant. (Thus, Bohr quantised toe angular momentum of the electron.).

The electron moving in any of the permitted orbits does not radiated energy in spite of its acceleration towards the centre of the orbit. The atom, therefore, is said to exist in a *stationary* state.

(ii) The emission (or absorption) of radiation by the atom takes place when an electron jumps for one permitted orbit to another. The radiation is emitted (or absorbed) as a single quantum (photon) whose energy hv is equal to the difference in energies of the electron in the two orbits involved. Thus, if E_i be the energy of the initial orbit of the electron and Z_f that of the final orbit then we have

$$hv = E_i - E_f,$$

where v is the frequency of the emitted (or absorbed) radiation.

Let e, m and v be the charge, mass and velocity of the electron (measured in column, kg and meter/sec respectively) and r the radius of the orbit measured in meter. The positive charge on the nucleus is Z e, where Z is the atomic number Fig. 5. In case of hydrogen Z = 1, so that positive charge on the nucleus is e. As the centripetal force is provided by the electrostatic attraction, we have

$$\frac{mv^2}{r} = \frac{1}{4\pi\varepsilon_0}\frac{e^2}{r^2}$$

or $$m\,v^2 = \frac{e^2}{4\pi\varepsilon_0 r}. \quad ...(1)$$

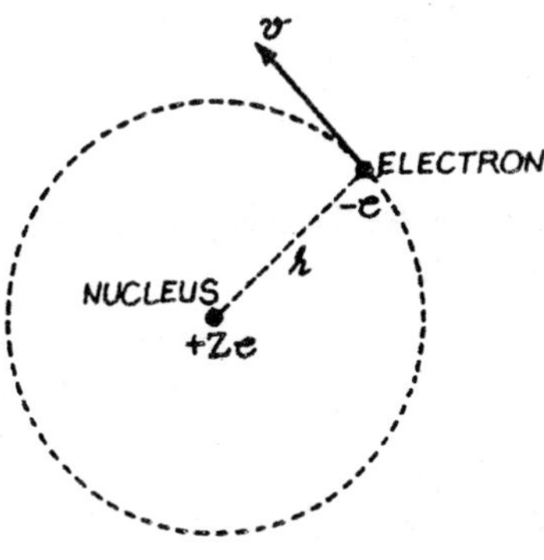

Fig. 5

From the first postulate, the angular momentum of the electron is given by

$$L = m\,v\,r = n\frac{h}{2\pi}, \quad ...(2)$$

where n is called as 'quantum number' having values 1, 2, 3, ...

Squaring eq. (2) and dividing by eq. (1) we get

$$r = n^2\frac{h^2\varepsilon_0}{\pi m e^2},\ n = 1, 2, 3, ... \quad ...(3)$$

This is the expression for the radius of the permitted orbits.

Now, the energy E of the electron in an orbit is the sum of kinetic and potential energies. The kinetic energy of the electron is

$$K = \frac{1}{2}mv^2 = \frac{e^2}{8\pi\varepsilon_0 r}. \quad \text{[from eq. (i)]}$$

The potential energy at a distance r to infinity against the electrostatic attraction $\left(-\frac{e^2}{4\pi\varepsilon_0 r^2}\right)$, and is given by

$$U = \int_r^{\infty} -\frac{e^2}{4\pi\varepsilon_0 r^2}\,dr = \frac{1}{4\pi\varepsilon_0}\left[\frac{e^2}{r}\right]_r^{\infty} = -\frac{e^2}{4\pi\varepsilon_0 r}.$$

Hence the total energy of the electron is

$$E = K + U = \frac{e^2}{8\pi\varepsilon_0 r} - \frac{e^2}{4\pi\varepsilon_0 r} = -\frac{e^2}{8\pi\varepsilon_0 r}.$$

Substituting for r from eq. (3), we get

$$E = -\frac{me^4}{8\varepsilon_0{}^2 h^2}\left(\frac{1}{n^2}\right) \quad n = 1, 2, 3... \qquad ...(4)$$

This is the expression for the energy of the electron in the n th orbit. We see that it in negative.

Let E_i and E_f be the energies of the electron corresponding to the initial (higher) and final (lower) orbits of the excited atom. Then, we have

$$E_i = \frac{me^4}{8\varepsilon_0{}^2 h^2}\left(\frac{1}{n_i{}^2}\right)$$

and $$E_f = -\frac{me^4}{8\varepsilon_0{}^2 h^2}\left(\frac{1}{n_f{}^2}\right),$$

where n_i and n_f are the corresponding quantum numbers. The energy difference between these states is

$$E_i - E_f = \frac{me^4}{8\varepsilon_0{}^2 h^2}\left(\frac{1}{n_f{}^2} - \frac{1}{n_i{}^2}\right).$$

Hence, from Bohr's second postulate, the frequency v of the emitted photon is

$$v = \frac{E_i - E_f}{h}.$$

$$= \frac{me^4}{8\varepsilon_0{}^2 h^2}\left(\frac{1}{n_f{}^2} - \frac{1}{n_i{}^2}\right)$$

The corresponding wavelength l is given by

$$\frac{1}{\lambda} = \frac{v}{c} = \frac{me^4}{8\varepsilon_0{}^2ch^3}\left(\frac{1}{n_f{}^2} - \frac{1}{n_i{}^2}\right).$$

This equation indicates that, *Since ni and nf can take only integral values, the radiation emitted by excited hydrogen atoms should contain certain discarte wavelengths only.*

The value of the constant *term* $\frac{me^4}{8\varepsilon_0{}^2ch^3}$ comes out to be the same as the Rydberg constant R in the Balmer's empirical formula. Thus we have

$$\frac{1}{\lambda} = R\left(\frac{1}{n_f{}^2} - \frac{1}{n_i{}^2}\right).$$

Emission of Spectrum

When the hydrogen atom gets sufficient energy from outside by some means, the electron from an inner orbit of lower energy goes up to an outer orbit of higher energy. This excited state of the atom lasts for jumping down, it emits the difference in energy between the two orbits as electromagentic radiation. If the electron jumps from an orbit ni to an orbit nf, the wavelength of the emitted radiation will be

$$\frac{1}{\lambda} = R\left(\frac{1}{n_f{}^2} - \frac{1}{n_i{}^2}\right).$$

It is found that for

$n_f = 1$,	$n_i = 2, 3, 4, ...$	we obtain	Lyman series,
$n_f = 2$,	$n_i = 3, 4, 5,...$	we obtain	Balmer series,
$n_f = 3$,	$n_i = 4, 5, 6$	we obtain	Paschen series,
$n_f = 4$,	$n_i = 5, 6, 7,...$	we obtain	Brackett series,
$n_f = 5$,	$n_i = 6, 7, 8,...$	we obtain	Pfund series,

The corresponding energy level diagram is shown in Fig. 6. The top horizontal line represents zero energy, that is, energy of the electron outside the atom ($n = \infty$). The other horizontal lines represent energies of different orbits given by the formula

$$E = -\frac{me^4}{8\varepsilon_0{}^2h^2}\left(\frac{1}{n^2}\right).$$

The arrows ending at the lines n = 1, 2, 3, 4, and 5 represent the transitions responsible for the Lyman, Balmer, Paschen, Brackett and Pfund series respectively.

PRINCIPLES OF THE BOHR'S THEORY

Bohr's theory although very successful l in explaining the hydrogen spectrum and giving valuable information about atomic structure, has the following shortcomings.

(i) An individual line of hydrogen spectrum, when examined under a high resolving spectroscope, is found to be accompanied by a number of theory as such. It can, however, be explained when the relativistic variation in the mass of the electron and the electron 'spin' are taken into account.

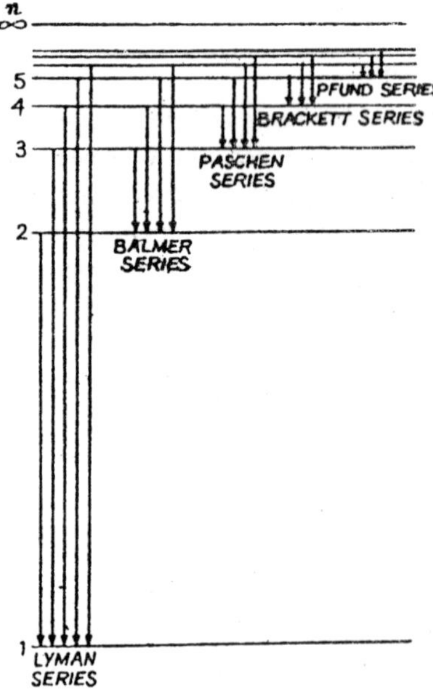

Fig. 6

(ii) Bohr's theory cannot explain the variation in intensity of the spectral lines of an element. The intensity can be explained by quantum mechanics.

(iii) The theory is only applicable to one-electron atoms such as hydrogen isotopes, singly-ionized helium, doubly-ionised lithium, etc. It does not explain the spectra of complex atoms.

(iv) The success of Bohr's theory in explaining the effect of magnetic field on spectral lines is only partial. The theory cannot explain the 'anomalous' Zeeman .effect.

(v) The theory does not satisfactorily explain the distribution of electrons in atoms.

NO BALMER LINES IN ABSORPTION SPECTRUM OF HYDROGEN

The Bohr theory also explains the absorption line spectrum of hydrogen. When a beam of - continuous light (containing all wavelengths) is passed through hydrogen and then sent into a spectrograph, a set of dark lines is obtained. In terms of quantum theory the incident light is a beam of quanta (photons) of all sorts of energies. Now according to Bohr theory the hydrogen atoms absorb only those quanta whose energies correspond to transitions between its discrete energy levels.

The resulting excited hydrogen atoms re-radiate the absorbed energy almost atones but these photons come off in random directions with only a few in the same direction as the original beam of continuous light. The dark lines in the absorption spectrum are therefore never completely black. Obviously the absorption lines will have exactly the same frequencies as the emission lines.

Now it is found that all the emission lines of hydrogen spectrum do not appear in the absorption spectrum. The reason is that normally the atom is always in the ground state $n = 1$. Therefore, absorption transitions can only occur from $n = 1$ to $n > 1$. Hence lines of only the Lyman series can appear in absorption spectrum.

To obtain Balmer series in absorption, the atom must initially be in the state $n = 2$, because Balmer lines require transitions from $n = 2$ to $n > 2$. Since atoms are usually in the ground state. Balmer lines are not obtained in absorption.

The simplest atomic spectrum is that of hydrogen. Its visible part consists of a single series which was first observed by Balmer in 1885. This is called the 'Balmer series' of hydrogen. Its first line having longest wavelength of 6563 Å (Fig. 7), is named Ha, the next Hb, and so on, the series limit reaching at 3646 Å (Fig. 7) Besides this, there is a series of lines in the ultraviolet part of the hydrogen spectrum which is known as 'Lyman series, and three series in the infra-red part which are known as 'Paschan series', 'brackett series' and 'Pfund series'.

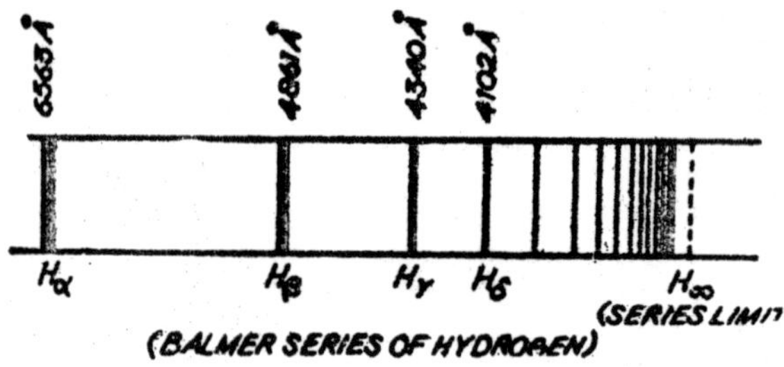

Fig. 7

Balmer discovered a formula for the wavelengths of all the lines of the Balmer series. His formula is

$$\lambda = 3646 \frac{n^2}{n^2 - 4} \text{ Å}, \; n = 3, 4, 5, \ldots$$

n = 3 gives the wavelength of H_α line, n = 4 gives H_β line, ... and n = ¥ gives the series limit.

Subsequently, Rydberg found that Balmer's formula was a special case of a more general formula which is as follows.

$$\bar{v} = \frac{1}{\lambda} = R_H \left(\frac{1}{m^2} - \frac{1}{n^2} \right),$$

where $\bar{v}$ is wave number (reciprocal of wavelength), m and n are positive integers (n > m) and RH is the Rydberg constant for hydrogen. Its value is 1.097 × 107 m^{-1}. To obtain formula for Lyman series we set m = 1 and n = 2, 3, 4, ..., for Balmer series m = 2 and m = 3, 4, 5, ... and so on. Thus.

$$\bar{v} = R_H \left(\frac{1}{1^2} - \frac{1}{n^2} \right), \; n = 2, 3, 4, \ldots \qquad \text{(Lyman)}$$

$$\bar{v} = R_H \left(\frac{1}{2^2} - \frac{1}{n^2} \right), \; n = 3, 4, 5, \ldots \qquad \text{(Balmer)}$$

$$\bar{v} = R_H \left(\frac{1}{3^2} - \frac{1}{n^2} \right), \; n = 4, 5, 6, \ldots \qquad \text{(Paschen)}$$

$$\bar{v} = R_H \left(\frac{1}{4^2} - \frac{1}{n^2} \right), \; n = 5, 6, 7, \ldots \qquad \text{(Brackett)}$$

$$\bar{v} = R_H \left(\frac{1}{5^2} - \frac{1}{n^2} \right), \; n = 6, 7, 8 \ldots \qquad \text{(Pfund)}$$

We see that the wave numbers of hydrogen lines can be expressed as differences of two terms of the form $\frac{R_H}{n^2}$.

The next simplest spectra are of the 'monovalent' atoms of alkali metals Li, Na, K, etc. The lines in the spectrum of an alkali atom can be grouped unto four distinct series, a principal series' of intense lines a 'sharp series' of fine lines, a diffuse series' of comparatively broader lines and a 'fundamental series' which lies in the infra-red region.

The sharp and diffuse series line in the visible part and converge to a common limit. Rydberg represented the lines of a particular series by the formula.

$$\bar{v} = Z^2 R_A \left[\frac{1}{(m - \Delta_1)} - \frac{1}{(n - \Delta_2)^2} \right],$$

where R_A is the Rydberg constant for a particular element A, Z is the atomic number, m and n are positive integers, and Δ_1 and Δ_2 are constants for the particular series. Actually each line of an alkali spectrum is a close doublet. Again, we find that the wave numbers of the lines can be expressed as differences of two terms like $\frac{Z^2 R_A}{(n - \Delta)^2}$.

After alkali spectra, next in complexity are the spectra of 'divalent' atoms of alkaline-earths Be, Mg, Ca. In a typical alkaline-earth spectrum, we can distinguish two distinct systems of lines, a system of singlets and a system of distinguish two distinct systems of lines, a system of singlets and a system of triplets. Each system has a principal series, a sharp series, a diffuse series and a fundamental series.

As we proceed to atoms having several valence electrons, the spectra becomes more complex and the groupings of lines into series becomes less pronounced. Still regularities can be observed in complex spectra, and it is possible to express the wave number of any spectral line as the difference of two terms.

RYDBERG-RITZ COMBINATION PRINCIPLE

The principle states *the wave numbers of spectral lines can be expressed by the difference of spectroscopic terms in such a way that other difference of those terms give also the wave numbers of lines in the same spectrum.* Suppose in a spectrum the wave numbers of two lines are given as

$$\bar{v}_a = T_2 - T_3 \text{ and } \bar{v}_d = T_1 - T_4,$$

Then lines of the following wave numbers are also expected in the same spectrum,

$$\bar{v}_b = T_2 - T4 \text{ and } \bar{v}_c = T_1 - T_3.$$

This means that constant differences exist between the wave numbers.

$$\bar{v}_b - \bar{v}_a = \bar{v}_d - \bar{v}_c$$

$$\bar{v}_c - \bar{v}_a = \bar{v}_d - \bar{v}_b.$$

DISPLACEMENT LAW

According to this law *the spectrum of any neutral atom of atomic number Z closely resembles the spectrum of the singly ionised atom of atomic number Z + 1.* For example, the spectrum of H^+ (Z = 2) closely resembles the spectrum of H (Z = 1) and can be represented by similar formula,

$$\bar{v} = 4\, R_{He} \left[\frac{1}{m^2} - \frac{1}{n^2}\right].$$

Other elements deprived of all but one electron, also produce hydrogen-like spectra which an be represented by the general formula

$$\bar{v} = Z^2 R_A \left[\frac{1}{m^2} - \frac{1}{n^2}\right].$$

In a similar way, the spectrum of a singly ionised alkaline-earth atom resembles the spectrum of an alkali atom. From this we conclude that it is the number of valence electrons in an atom which determines the qualitative character of the spectrum of that atom.

Hydrogen Spectrum

The spectrum of hydrogen atom consists of a number of lines. These lines have been grouped into a number of 'series'. The lines in each series are such that *their separation and intensity decrease regularly towards shorter wavelengths,* converging to a limit called the 'series'limit. The wavelengths in each series can be given by a simple empirical formula. The first such spectral series was observed by balmer in 1885 and is called the Balmer series of hydrogen. The first line with the longest wavelength (6563 Å) is named Ha. The next Hb, and so on. The series limit lies at 3646 Å, beyond which is a faint continuous spectrum. Balmer's formula for the wavelengths of the series is

$$\frac{1}{\lambda} = R\left(\frac{1}{2^2} - \frac{1}{n^2}\right), \ n = 3, 4, 5, \ldots \qquad \text{(Balmer)}$$

The quantity R is called the 'Rydberg constant' and has the value

$$R = 1.097 \times 10^7 \text{ meter}^{-1}.$$

The Ha line corresponds to n = 3, the H_β line to n = 4, and so on. The series limit experiment. The Balmer series contains only those spectral lines which fall in the visible part of the hydrogen spectrum. The lines falling in the ultraviolet and infrared parts form other series. The lines in the ultraviolet form the Lyman series whose wavelengths are give by

$$\frac{1}{\lambda} = R\left(\frac{1}{1^2} - \frac{1}{n^2}\right), \ n = 2, 3, 4, \ldots \qquad \text{(Lyman)}$$

In the infrared, three spectral series have been observed whose lines have the wavelengths given by the formulas

$$\frac{1}{\lambda} = R\left(\frac{1}{3^2} - \frac{1}{n^2}\right), \ n = 3, 5, 6, \ldots \qquad \text{(Paschen)}$$

$$\frac{1}{\lambda} = R\left(\frac{1}{4^2} - \frac{1}{n^2}\right), \ n = 5, 6, 7, \ldots \qquad \text{(Brackett)}$$

$$\frac{1}{\lambda} = R\left(\frac{1}{5^2} - \frac{1}{n^2}\right), \ n = 6, 7, 8, \ldots \qquad \text{(Pfund)}$$

Different Spectral Lines of Hydrogen

Every atom when excited emits radiations. The radiations form a line spectrum which is the characteristic of the emitter. Each atom has its own particular line spectrum which is regarded as the characteristic of that element to which the atom belongs. Like other elements, hydrogen possesses its own characteristic line spectrum. Hydrogen spectrum consists of a number of lines. These have been grouped into five series which are named after their discoverers, Fig 8. Many attempts were made by various workers to find a rule for underlying relationship which governed the wave lengths of these lines. We will discuss these by one.

Fig. 8

Calculation of Rydberg's Constant

The energy of the hydrogen atom when the electron is in the n_1th orbit.

$$En_1 = -\frac{me^4}{8\varepsilon_0{}^2h^2}\cdot\frac{1}{n_1{}^2}Z^2$$

and atom the energy of the atom when the electron is in the n_2th orbit

$$En_2 = -\frac{me^4}{8\varepsilon_0{}^2h^2}\cdot\frac{1}{n_2{}^2}Z^2$$

and the frequency of the photon emitted, when an electron jumps from n_2th to n_2th orbit is given by Bohr's third postulate, *i.e.*,

$$hv = En_1 - En_2 = -\frac{me^4}{8\varepsilon_0{}^2h^2}\left(\frac{1}{n_1{}^2}-\frac{1}{n_2{}^2}\right)Z^3$$

or

$$v = \frac{me^4}{8\varepsilon_0{}^2h^2}\left(\frac{1}{n_2{}^2}-\frac{1}{n_1{}^2}\right)Z^2.$$

Since the velocity of light c = vλ, we have

$$\frac{v}{c} = \frac{1}{\lambda} = \frac{me^4}{8\varepsilon_0{}^2ch^2}\left(\frac{1}{n_2{}^2}-\frac{1}{n_1{}^2}\right)Z^2 \qquad ...(1)$$

where

$$R = \frac{me^4}{8\varepsilon_0{}^2ch^2}.$$

On substituting the values, we get R = 109737.302 cm^{-1}. Here ε_0 is a constant whose numerical value is equal to 8.854. $\times 10^{-12}$. This value is found to be in agreement with the value obtained from the spectroscopic data of the Balmer's series which is 109677.67cm^{-1}.

Success of Bohr's Theory : This theory explains the following facts about atomic spectra.

Failures of Bohr's Theory

(i) It failed to explain *Stark effect.* It is similar to Zeeman effect and is produced in the presence of external electrostatic field (Fig. 5).

(ii) It failed to explain Zeeman effect. When a substance emitting a line space trum is placed in a magnetic field, its lines would split up into a number of closely spaced lines. This is known as *Zeeman effect* (Fig. 9)

(iii) When spectral lines of hydrogen are observed very closely, each line is further made up of much closely spaced lines.

(iv) It failed to explain spectra of atoms other than hydrogen.

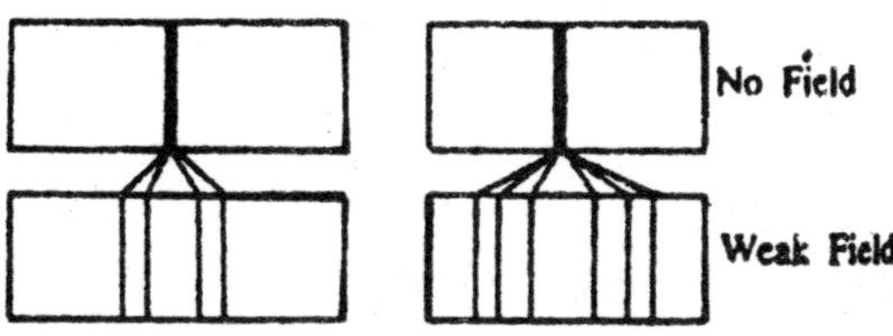

Fig. 9

(v) Another fundamental objection against Bohr's theory is that is used two theories which are opposed to each other, *i.e.*, *quantum theory* was used to account for the existence of stationary orbits and for frequencies of radiations emitted while motion of electron in its orbit obeyed *the law of classical mechanics.*

(vi) In the light of the Heisenberg's uncertainty principle, both the velocity and the position of the electron cannot be specified at a given time as Bohr did. Thus, it is another weakness of Bohr's theory.

(vii) Another weakness of Bohr's theory is that it did not throw light on the distribution and arrangement of electrons in atoms.

(viii) Bohr assumed that the nucleus is stationary and only electrons are revolving around it. Detailed facts have revealed that both the nucleus and the electrons move in closed orbits around their centre of mass. Therefore, it is major weakness of Bohr's theory.

Light Phase

Evidence in support of light phase. Hill and *Scrisbrick* found that isolated chloroplasts when illuminated produced oxygen from water. If certain hydrogen acceptors (oxidants) were present in the medium, they were reduced by the chloroplasts. The oxidants are called *Hill reagents* and the reaction as *Hill reaction.* Hill attributed the production of oxygen and hydrogen from water by light. *Arnonin* 1954, working on isolated chloroplasts reported that in addition to carrying out the Hill reaction, the chloroplasts could also synthesis ATP in the light from ADP and inorganic phosphate (Pt). This *photosynthetic phosphorylation* may be represented as.

$$ADP + P_t \xrightarrow{\text{Light energy}} ATP.$$

Reaction occurring in Light phase.

When chlorophyll molecule absorbs one quantum of light, it gets excited. The excited chlorophyll molecule brings about the following changes.

(i) Excited chlorophyll molecule interacts chemically with water to liberate oxygen.

$$4H_2O + \text{chlorophyll}^* \rightarrow 4[H] + 4[OH] + \text{chlorophyll}$$

$$4[OH] \rightarrow 2H_2O + O_2.$$

The nascent hydrogen reacts with NADP to form $NADPH_2$ which acts as a reducing agent.

$$NADP + 2H \rightarrow NADPH_2$$

(ii) Excited chlorophyll also brings about the reaction between ADP and inorganic phosphate Pt to form an energetic compound ATP.

$$ADP + P_i + \text{chlorophyll}^* \rightarrow ATP + \text{chlorophyll}.$$

The products obtained from light reaction enter the dark reaction of photosynthesis.

Mechanism of light phase reactions. When sun light falls on the leaves, the chlorophyll present in the leaves absorbs light photon and becomes excited. This photo excitation results in the displacement of an electron from the normal orbit (of chlorophyll) into a anew orbit. The hole left by the displaced electron is soon filled up by the return of the same electron or another one. If the same electron is returning the process is termed as *cyclic-transfer of electron.* However, if the different electron is returning, it is termed as non-cyclic transfer of electrons.

$$\text{Chlorophyll} + h\nu \rightarrow \text{Chlorophyll}^*$$

$$\text{Chlorophyll}^* \rightarrow (\text{chlorophyll})^* + e^-$$

Non-cyclic electron transfer. It involves the following steps.

(i) First of alkyl water dissociates into hydrogen and hydroxyl ions.

$$4H_2O \rightarrow 4H^+ + 4OH^-$$

(ii) When a quantum of short wavelength of light is received by a pigment II of chlorophyll molecule, it loses an electron. The loss of electrons is immediately compensated by the electrons of 4 OH^- ions, *i.e.,* converted into water and oxygen.

$$4\ OH^- \rightarrow 2H_3O + O_2 + 4e^-$$

(iii) The electron released from above process is raised from + 0.8 eV to zero potential at which it is captured by an electron carrier called *Plastoquinone.*

(iv) After this the electron travels downhill and falls back to + 0.4 eV in a dark reaction through a series of carriers (cytochrome b_6, cyt. f, and plastocyanine) to pigment system I. The energy released in the dowahill passage of electron from cytochrome b_6 to cytochrome f is utilised to convert ADP and inorganic phosphate into ATP.

$$ADP + P_i \rightarrow ATP$$

(v) The electron is pumped from + 0.04 to – 0.4 eV with energy of another quantum absorbed by the system. I. According to *Tagawa* and *Arnon* (1962) the electron from system I is captured by any iron-containing protein, called *ferredoxin* with a potential of – 4.32 eV.

(vi) According to *Aron* (1967) the electron is finally passed on at – 0.32 e.v. to NAD which together with two H+ released from water, becomes reduced to NHDPH + H^+ (expressed as $NADPH_2$ for convenience).

$$NADP^+ + 2e^+ + 2H^+ \rightarrow NADPH + H^+$$

Energy from ATP helps to more the electron from $NADPP_2$ to phosphoglyceric acid into carbon cycle.

Cyclic electron transfer. The expelled electron from the chlorophyll molecule is first accepted by *ferredoxin* or by an other electron acceptor of the chloroplast.

The electron then traverses via cytochrome b_6 and cytochrome f, the two native cytochromes of chloroplasts, and ultimately reaches the original chlorophyll molecule from which it has been expelled. This scheme of electron transport is called *cyclic photophosphorylation.* In the cyclic photophosphorylation the electron, that is returned to the chlorophyll, is the same as the one that was expelled. In cyclic photophosphorylation energy is released during the transfer of electrons to cytochromes and the released energy is utilized to synthesize ATP.

Dark Phase

The present knowledge on the path of carbon in photosynthesis comes mainly from the work of *Melvin Calvin* and his associates,. They allowed the algae supplied with carbonlabelled CO_2 to photosynthesize for very brief periods of time and killed them instantaneously with boiling alcohol. Such studies using tracer technique revealed that products of CO_2 assimilation even for periods less than a minute were mainly sugars and amino acids shown in Fig. 10.

Calvin's group was interested to know the first stable product of photosynthesis. For this they allowed the algae to photosynthesis for 5 second and extracted the compounds after killing the organisms. The first stable product of very brief period of photosynthesis was found to be 3 carbon compound phosphoglyceric acid (PGA). Since PGA is a 3 carbon compound, it was thought that CO_2 is accepted by a 2-carbon molecule to form 3 carbon PGA. But no such compound was detectable in plants.

Later, *Benson* working in the same laboratory found that a 5-carbon sugar Ribulose diphosphate (RuDP) is the acceptor molecule of CO_2. On accepting CO_2 it was established that an unstable carbon intermediate is found and it cleaves into 2 molecules of PGA.

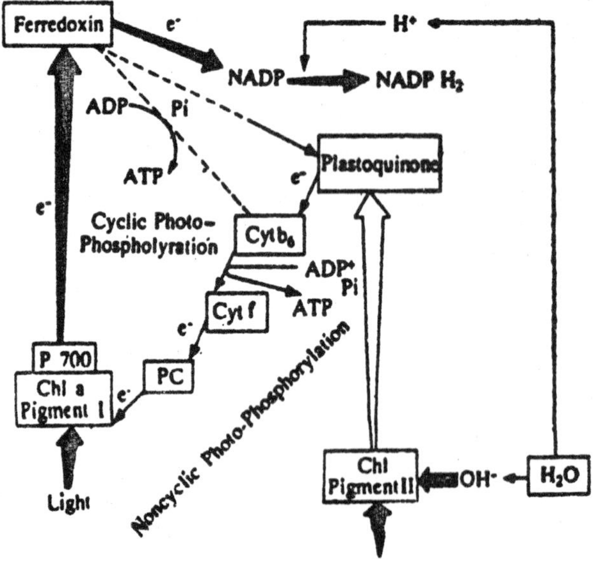

Fig. 10

The end products of photosynthesis, namely sugars or starches, are synthesized from PGA by reversed sequence of EMP pathways of glycolysis. It was also established that some of the PGA is used to regenerate RuDP by accepting some more CO_2 molecules. Thus, there is a regular cycle of reactions that ultimately produce the end product of photosynthesis.

The formation of hexose sugar in the dark reaction may be summarised as follows.

(a) 3 molecules of carbondioxide combine with 3 molecules of RuDP (Ribulose diphosphate) to form 6 molecules of phosphoglyceric acid (PGA).

$$3CO_2 + 3RuDP \rightarrow 6PGA$$

(b) 6 molecules of phosphoglyceric acid (PGA) combine with 6 molecules of ATP, which are formed in the light reaction, to form 6 molecules of diphosphoglyceric acid (DPGA).

$$6PGA + 6ATP \rightarrow DPGA + 5\ ADP$$

(c) 6 molecules of DPGA are reduced to 6 molecules of phosphoglyceraldehyde by 6 molecules of $NADPH_2$ which are produced in light reaction.

$$6\ DPGA + 6\ NADPH_2 \rightarrow 6\ \text{phosphoglyceraldehyde}\ H_2PO_4 + 6\ NADP$$

(d) Now one of the molecules of phosphoglyceraldehyde is utilised in the formation of a half molecule of hexose.

$$1\ \text{Phosphoglyceraldehyde} \rightarrow 1/2\ (\text{hexose}).$$

The five other molecules of phosphoglyceraldehyde are utilised for the continued regeneration of ribulose diphosphate.

$$5\ \text{Phosphoglyceraldehyde} \xrightarrow{\text{Changes}} 3\ \text{Ribulose Diphosphate.}$$

(i) The first molecule of phosphoglyceraldehyde is converted into its isomer, dihydroxy acetone phosphate. The latter molecule combines with second molecule of phosphoglyceraldehyde to form fructose 1, 6 diphosphate which in its turn gets dephosphorylated to form fructose monophosphate.

$$\text{Phosphoglyceraldehyde} \xrightarrow{\text{Isomerises}} \text{Dihydroxyacetone phosphate}$$

$$\text{Dihydroxy acetone phosphate} + \text{phosphoglyceraldehyde} \longrightarrow$$

(II)

$$\text{Fructose, 6 monophosphate} \xrightarrow{ADP \rightarrow ATP} \text{Fructose 1, 6 diphosphate.}$$

(ii) The fructose 6 monophosphate now combine with the third molecule of phosphoglyceraldehyde to give rise erthrose monophosphate (C_4) and xyluose monophosphate (C_5).

Fructose – 6 – monophosphate + Phosphoglyceraldehyde
(III)
⟶ Erythrose monophosphate (C_4)
+ Xylulose monophosphate (C_5)

(iii) The erythrose monophosphate now combines with the fourth molecule of phosphoglyceraldehyde to form sedoheptulose diphosphate which loses one phosphate to become sedoheptulose monophosphate. It then combines with the fifth molecule of phosphoglyceraldehyde to give rise to one molecule each of xylulose monophosphate and ribose monophosphate.

Erythrose monophosphate + Phosphoglyceraldehyde.
⟶ sedoheptulose diphosphate
↓ ADP – ATP

Xylulose monophosphate + Ribose monophosphate $\xleftarrow[\text{(V)}]{\text{Phosphoglyceraldehyde}}$ Sedoheptulose monophosphate

The ribose monophosphate gives rise to one molecule of ribulose monophosphate. The two molecule of xylulose monophosphate, one from this step and another form step (ii), are reversibly converted into its isomer ribulose monophosphate.

In this way five molecules of phosphoglyceraldehyde are converted into three molecules of ribulose monophosphate which in turn gets phosphorylated to form ribulose diphosphate. Now by doubling the number of CO_2 molecules of six, the path of carbon in dark phase of photosynthesis.

Explanation of the Doublet Structure

When we study the spectra of elements other than hydrogen, we observe that many lines are actually multiples consisting of two, three or more lines close together.

In order to explain Uhlenbeck and Goldsmit (9135) suggested that this multiplicity of spectral lines is due to the spin of the electron. The contribution due to the spin is quantised and is expressed in terms of spin quantum numbers. The value of this number is + 1/2 and – 1/2. The resultant of the spin and azimuthal quantum number is given by j = l + s, where 'j' is called the *inner quantum number.* As 's' can have + 1/2 and + 1/2, it follows that for every value of 'l' there are two values of 'j' viz.,

$$j_1 = l + 1/2 \text{ and } j2 = l - 1/2.$$

The above values hold good except when l is zero. In that case the two 'j' values are identical, *i.e.*, + 1/2, because it is not the + ve or – ve sign but is the numerical value only of 'j' which determines the momentum. It means that every 'l' level except l = 0 is consequently split into two levels with different energies. Let us now illustrate this discussion by applying to alkali metals.

(i) For an electron in an 'S' level, the value of l = 0. It means that the two values of j are numerically identical and hence it is a singlet level.

$$j_1 = l + \frac{1}{2} \text{ and } j_2 = l - \frac{1}{2}$$

$$j_1 = 0 + \frac{1}{2} \text{ and } j_2 = 0 - \frac{1}{2}$$

$$j_1 = + \frac{1}{2} \text{ and } j_2 = - \frac{1}{2}$$

(ii) In the P level, l = 1 and so the corresponding values of 'j' are 3/2 and 1/2.

$$j_1 = l + \frac{1}{2} \text{ and } j_2 = l - \frac{1}{2}$$

$$j_1 = 1 + \frac{1}{2} \text{ and } j_2 = 1 - \frac{1}{2}$$

$$j_1 = \frac{3}{2}\ j_2 = \frac{1}{2}.$$

Thus, each spectral line in 'P' level is doublet.

Similarly in the D and F levels, the values of 'j' are 5/2 and 3/2, 7/2 respectively. Again 'j' has two values and therefore each line will be a doublet. The doublet of the sharp series of spectra may be represented by the expressions

$$\bar{v} = aP_{1/2} - nS$$

and

$$\bar{v} = aP_{3/2} - nS$$

where S is a singlet and P is a doublet.

Photochemical Inhibition

There are certain substances which are able to retard the rate of a photochemical reaction when present in trace amounts. Such substances are known as *inhibitors* and the phenomenon is known as *photochemical inhibition.* Some examples are.

(i) Traces of nitric oxide and propylene lower the quantum yield of the photochemical combination of hydrogen and chlorine.

(ii) Traces of impurities like NH_3 which when present in the hydrogen and chlorine reaction lower its quantum yield from 10^6 to 10^4. In 1905, Chapman showed that the inhibitor action of NH_3 is due to the side reaction of NCl_3 which may result due to the reaction of NH_3 with Cl_2.

$$NCl_3 + 3HCl \rightleftharpoons NH_3 + 3Cl_3$$

$$NCl_3 + 3H_2O \rightleftharpoons NH_3 + 3HClO$$

At equilibrium the concentration of NH_3 is maintained and the actual formation of HCl is reduced,

(iii) SO_2 and O_2 were also found to be inhibitors in various photochemical reactions.

Explanation. It is generally accepted that photo-inhibitors interrupt the chain reactions by removing chain carrier atoms or radicals.

Period of Induction

Many reactions are characterised by an initial period during which the process appears to be silent. This time is known as period of induction. Some examples are.

(i) An induction period is observed in hydrogen and chlorine reaction when certain impurities are present. With pure hydrogen and chlorine, no induction period is observed.

(ii) An induction period is also observed in the absorption of As_2O_2 by mercuric oxide in the presence of light.

Period of induction is due to presence of impurities like ammonia. These substances act as inhibitors by breaking these chains. When these impurities are completely converted into nitrogen and ammonium chloride during the reaction, then only normal photochemical reaction can take place.

THE THOMSON'S MODEL

Rutherford's experiment of radioactivity proved the facts that atom consists of positively and negatively charged particles. The question arose for the equilibrium of the atoms. Thomson suggested that the atom was spherical in shape. The whole mass of the atom was evenly distributed and the positive charge was distributed all over the mass. The electrons with negative charge were embedded within the atom. The whole positive charge of the atom was equal to the total charge of the electrons. The atom was like a plum *pudding*. The electrons oscillate or vibrate about their mean positions.

The main feature's of Thomson Model were:

1. It gives a mechanism for the emission of the eletro-magnetic waves. The electro-magnetic waves were supposed to be emitted by the oscillating electrons in the atom.
2. It explained the emission of electrons by heating as thermions and in photo-electric effect.
3. It provided an elastic hard model required for the kinetic theory of gases.

PHOTOCHEMICAL EQUILIBRIUM

A state of photochemical equilibrium is said to exist in a reaction when the rates of two opposing reactions of which at least one is light sensitive, become equal under the influence of light radiation.

Types : A number of cases of the photostationary state have been studied. These cases fall into two categories.

(a) In the First Category, only one reaction is light sensitive

$$A + B \underset{\text{dark}}{\overset{\text{light}}{\rightleftharpoons}} C + D$$

Some examples of such type are.

(i) *Dissociation of Nitrogen Dioxide*. When the vapour of nitrogen dioxide is exposed to light radiation of wavelength less than 3700Å, it dissociates to form nitric oxide and oxygen but the combination of these products is a dark reaction. Hence the equilibrium is

$$2NO_2 \underset{\text{dark}}{\overset{\text{light}}{\rightleftharpoons}} 2NO + O_2$$

When the reaction just starts, the pressure of the gas rises due to the decomposition of the nitrogen dioxide but becomes constant as soon as the stationary state it reached.

Dimerisation of anthracene. Another example of a photo-stationary state is the dimerisation of anthracene in solution in which the forward process is light sensitive but the back ward reaction is a thermal reaction, and so the equilibrium is

$$2_{14}H_{10} \underset{\text{dark}}{\overset{\text{light}}{\rightleftharpoons}} C_{28}H_{20}$$

Let us calculate the equilibrium constant of the reaction of category (a),

light $$A + B \underset{\text{dark}}{\overset{\text{light}}{\rightleftharpoons}} C + D$$

As the forward reaction is light sensitive, its rate will not depend upon the concentrations of A and B but upon the intensity of absorbed light. Hence.

Rate of the forward reaction = $k_1 I_{abs}$ where k_1 is constant for the forward reaction. As the backward reaction is a dark or thermal reaction, its rate will depend upon the concentrations of products C and D, *i.e.*,

Rate of backward reaction = k_2 [C] [D].

At photostationary state.

Rate of forward reaction = Rate of backward reaction

or $$\frac{k_1}{k_2} = \frac{[C][D]}{I_{abs}} \quad \text{or} \quad K = \frac{[C][D]}{I_{abs}}$$

where K is the equilibrium constant and is defined as the ratio of velocity constants of two opposing reactions.

The equilibrium constants for photochemical equilibria are constant only for a given light intensity, and vary as the latter is charged. Photochemical equilibrium constant is generally independent of temperature changes if the intensity of light is kept constant.

(b) In the Second Category, both the reactions are light-sensitive.

$$A + B \underset{\text{light}}{\overset{\text{light}}{\rightleftharpoons}} C + D$$

Few examples of this category are

(i) *Formation of sulphur trioxide.* It is an example of a photostationary state in which both the forward and back ward processes are light sensitive.

$$2SO_2 + O_2 \underset{\text{light}}{\overset{\text{light}}{\rightleftharpoons}} 2SO_3 .$$

The value of photochemical equilibrium constant of this reaction is almost independent of temperature from 500° to 800°C.

(ii) *Isomerisation of maleic acid into fumaric acid.* Another example is the isomerisation of maleic acid into fumaric acid in which both forward and backward reactions are light sensitive.

$$\begin{array}{c} H—C—COOH \\ \| \\ H—C—COOH \\ \text{Maleic acid} \end{array} \underset{\text{light}}{\overset{\text{light}}{\rightleftharpoons}} \begin{array}{c} H—C—COOH \\ \| \\ HOOC—C—H \\ \text{Fumaric acid} \end{array}$$

The value of equilibrium constant can be calculated in the similar manner as given in category (a).

Application. The concept of photostationary state has been used to explain the phenomenon of *vision.* When the light sensitive substance in eye called visual purple exposed to light, the latter gets bleached to form visual yellow.

Thus, an equilibrium state is established between visual purple and visual yellow. In dark the visual purple accumulates and eye becomes sensitive. When exposed to light further, it results in the phenomenon of *dazzling.*

Bioluminescence

Certain living organisms emit light and show the phenomenon of chemiluminescence. It is known as bioluminescence. The phenomenon of bioluminescence was investigated by *E. N. Harvey* (1915). Some interesting examples are.

(i) The cold light produced by certain living organisms like fire-fly or glow worm is probably due to the oxidation of protein, luciferin, by the atmospheric oxygen in the presence of enzyme luciferase. The firefly emits light having a maximum intensity of wavelength of 5700Å. It can be easily judged by naked eye.

(ii) Certain marine animals also emit light. The protozoa Noctiluca of the sea produces phosphorescence. It has also been observed that many deep sea fishes have powerful organs which emit dazzling light.

PATH OF CARBON DIOXIDE IN PHOTOSYNTHESIS

Photosynthesis

The energy which supports the activities of most living organisms on the earth is derived directly or indirectly form the energy of sunlight through photosynthesis *Photo* = light, *synthesis* = to build up). *It is a process in which simple carbohydrates are synthesised from water and carbon dioxide in the chlorophyll containing tissues of plants in the presence of sunlight. Oxygen being a by-product is given out*

Kamen (1963) has defined photosynthesis as a *series of processes in which electromagnetic energy is converted into chemical free energy which can be used for biosynthesis.* By a single overall equation, it may be shown as below.

$$CO_2 + H_2O \xrightarrow[\text{chlorophyll}]{hv} 1/6\,(C_6H_{12}O_6) + O_2$$

In more general way, it is

$$CO_2 + H_2O \xrightarrow[\text{chlorophyll}]{hv} 1/n\,[CH_2O]_n + O_2$$

Scarce of Oxygen Liberated in Photosynthesis : The released oxygen (O_2) in photosynthesis comes from water [H_2O^{18}]. When green plants were supplied with water containing [H_2O^{18}]. The released oxygen was entirely of the O^{18} type, indicating that water is only source to release oxygen in photosynthesis. This can be represented by the following summary reaction.

$$6CO_2 + 12H_2O^{18} \rightarrow C_6H_{12}O_6 + 6H_2O + 6O_2^{18}.$$

Photosynthesis and Pigments. The photosynthetic products are energy-rich organic compounds. The potential chemical energy of these compounds comes from the light energy. The light' energy to be effective in photosynthesis must be absorbed by a suitable pigment. The vital role is performed by the green pigment chlorophyll, in plants.

Chlorophyll pigment. There are at least seven types of chlorophylls known : chlorophylls *a, b, c, d* and *e.* bacteriochlorophyll and bacterioviridin. All these chlorophyll molecules contain a *tetrapyrrole* skelton formed into a ring with an atom of magnesium in the centre of the ring. A so-called pyrrole molecule contains a skeleton of five atoms, four carbon and one nitrogen and five are arranged in a ring.

Four such pyrroles arranged in a ring form the 'head' of a chlorophyll molecule. Attached to this *porhpyrin* ring at one point is an alcohol (phytol) "tail", a long chain of linked carbons. Relatively minor variations in the kinds and groupings of other atoms joined to this head and tail skeleton account for the differences among different kinds of chlorophylls.

Chlorophylls *a* and *b* are the two most abundant ones found in all the autotrophic plants except the pigment containing bacteria. Chlorophyll *b* is however, absent in blue-green, brown, and red algae. The other chlorophylls (c, d, e) are found only in algae in combination with chlorophyll a. Chlorophyll *a* possesses a – CH_3, a methyl group which is replaced by a – CHO, an aldehyde group, in chlorophyll *b*. The air contains 0.03% carbon dioxide. This gas diffuses into the air spaces between the cells of the mesophyll of the leaf through the open stomata. Here it combines with water to form carbonic acid.

Transport of carbon dioxide in the form of carbonic acid takes place from the spongy cells to the palisade cells until the main site of photosynthesis is reached. It is now proved that in photosynthesis carbon dioxide is not reduced as such but it first forms a complex with cellular compounds and is then reduced by some 'activated' compounds produced as a result of photochemical reaction.

Since isolated choloroplasts are unable to reduce carbon dioxide in the light, it is concluded that either carbon dioxide uptake must take place outside the chloroplast or some mechanism has been destroyed during extraction of the chloroplasts. The first alternative is more probable. *Frenkel* (1941) prepared CO_2 from radioactive (C^{14}) carbon. He fed *Nitella* cells with this carbon dioxide ($C^{14}O_2$) in the dark and found that all the radioactive carbon was located in the cytoplasm. When exposed to light, 4/5 of radioactive carbon was found in the chloroplasts. The simplest interpretation of Frenkel's result is that carbon dioxide uptake occurs in the cytoplasm whereas reduction of carbon dioxide is associated with the chloroplasts.

MECHANISM OF PHOTOSYNTHESIS

Various theories have been propounded by various workers. But the earlier theories failed and still not theory is available which describes the mechanism satisfactorily. The uncertainty in the mechanism is die to.

(a) No products of the earlier stages of the process could be detected experimentally.

(b) When chlorophyll is extracted from plant, it fails to reproduce its photosynthetic property. So the photosynthesis cannot be studied under controlled experimental conditions.

(c) The complete structure of the chlorophyll was not known fully.

The mechanism of photosynthesis as currently understood is usually divided into three phases :

(1) The absorption of the light energy by the pigment system,

(2) Conversion of light energy into chemical energy by photo-phosphorylation, and

(3) Synthesis of organic compounds by the products of the photo-chemical reactions.

Enough evidence has accumulated in the recent past to show that photosynthesis consists of two stages, (a) light phase and a dark phase. The reaction of light phase is light sensitive and 'therefore' called a photochemical reaction. The reactions of the dark phase do not require light and are temperature sensitive. Such reactions are after called *Blackman reactions.*

$$\xrightarrow[\text{Photochemical reaction}]{\text{Step I}} \xrightarrow[\text{chemical reaction}]{\text{Step II}} \text{Photosynthetic Product}$$

(The component reactions of photosynthesis).

Evidences for the Existence of Light and Dark Phases : There are following evidences.

1. *Experiments with intermittent light.* When Chlorella cells were exposed to continuous illumination, *Warbarg* found that the rate of photosynthesis was less when compared to those exposed alternately to light and darkness. The increase in the rate of photosynthesis was enhanced greatly if the span of the light period was reduced. This is because light period followed by darkness allows the cells to utilize all the products of the light reaction for the reduction of CO_2 in the dark reaction. When the light was continuous or interrupted, the products of light phase accumulate because the pace of light reaction is far greater than that of dark.

2. *Temperature Experiments.* When the photosynthesis is carried out in the excess of carbon dioxide but in the presence of low light intensity, the rate of photosynthesis does not increase with the increase in temperature. This indicates that some reaction requires light and is not affected by rise in temperature. When the concentration of carbon dioxide is low and the intensity of

light is high, the rate of photosynthesis becomes double for every 10°C rise in temperature. This shows that there must be a reaction which is not affected by light. This is the dark reaction of photosynthesis.

Let us discuss these two stages separately.

Latent Image

The usefulness of present-day photographic materials is a result of the behaviour of the tiny individual particles of silver halide which are contained in photographic emulsion. In many cases these particles are extremely sensitive to light and store up the effect to an exceedingly small amount of light. The effect çan, in turn, be multiplied many fold by the action of a developer. *This stored up effect of light is known as the latent image.*

It is so small that it cannot be developed itself. Consequently, the nature of the latent image and the nature of the development process are closely related.

Def. *When a photographic plate is exposed to light radiation for a certain time, an invisible image is formed on it. This invisible image is known as latent image.*

Facts

(i) In the formation of latent image, no change in the emulsion has been detected.

(ii) The formation of latent image is very sensitive to traces of substances such as Ag, Ag_2 S etc.

(iii) When the photographic plate carrying latent image is exposed to mild reducing agent like pyrogallol, the latent image is developed to a negative plate.

Studies of the behaviour of the latent image under different conditions of development and exposure and its reactions with certain other chemicals have led to a fairly clear conception of its nature and the mechanism of its formation. An understanding of these processes requires knowledge of the structure of the silver halide grains themselves.

Theories of Latent Image Formation : When a film is exposed to light radiation, the absorbed energy ejects an electron from the bromide ion.

$$Br^- + hv \rightarrow Br + 1e^-.$$

The electron thus set free is captured by a silver ion to form a silver atom.

Thus, the overall reaction is the photochemical dissociation of silver bromide into silver and bromine by absorbing one quanta of light, *i.e.*,

$$AgBr + hv \rightarrow Ag + Br.$$

Thus during the formation of latent image, silver is formed and bromine atom being removed by gelatin.

Objection. The main problem is that light is absorbed all over the surface of the photographic film but silver atoms make their appearance only at a few isolated points within each grain to form a latent image. How does it happen ? How is the latent image formed ? Various theories regarding the formation of latent image are.

I. Paul's View : According to Paul there are several empty holes in the crystal. When a positive ion is missing at one point a negative ion will be missing at some other point. Thus, there exists some holes in the crystal.

When light is incident on a photographic plate, the electrons are given out and finally fall inside the holes. Thus,

$$X^- + hv \rightarrow \quad X + e^-$$

the latent image was formed at these holes. When the exposed photographic plate is developed, negative image is formed at these holes. This theory does not explain the formation of latent image in complete manner.

+	−	+	−	+	−
−	+	−	+	−	+
+	−	hole	−	+	−
−	+	−	+	−	+
+	−	+	−	+	−
−	+	−	hole	−	+
+	−	+	−	+	−
−	+	−	+	−	+

Fig. 11 : Formation of latent image.

II. Webb's Theory : Recent advances in quantum mechanics and the applications of these principles to the structure and behaviour of crystals have contributed a great deal to knowledge of the problem of latent image formation. Webb (1936) first applied the principles of quantum mechanics to an explanation of the formation of the latent image.

Webb described the latent image as a concentration of electrons at a speck, or trap. The electrons are supposed to be transferred to a higher energy level by the absorption of light and in this energy level they could wander through the crystal at will as in the photo-conductance effect. The sensitivity specks were supposed to contain energy levels slightly lower than this conductances level, so that when an electron travels to such a specks, it gives up a bit of energy in going to this new level and is trapped. This gave an excellent explanation of the speck concentration of the effect of absorbed light but did not explain the formation of metallic silver as latent image substance.

III. Gurney and Mott's Theory : Gurney and Mott presented a theory for the formation of latent image.

1. The energy structure of a crystal of silver halide can be described as in Fig. 12. The Ag and Br represent the silver and bromide ions which are present alternately in a crystal. The crosshatches labelled as P represent the energy level of the electrons connected with the bromide ions. This energy band is filled and electrons cannot move around in it. The clear energy band labelled S represents the energy level corresponding to the conductance level in metallic silver. In a crystal of silver halide all the valence electrons are associated with the bromide ions and none with silver so that this band is empty.

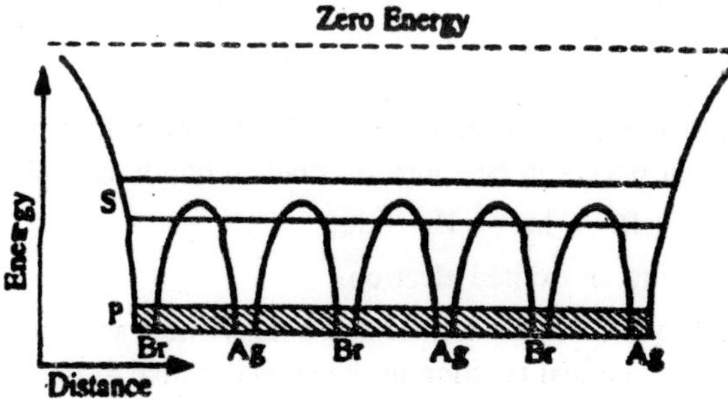

Fig. 12 : Quantum mechanical description of the energy level occurring in a crystal of silver bromide.

Electrons cannot have energy represented by the areas not included by these two bands. The photo-conductance phenomenon can be explained by this diagram–when light is absorbed, its energy is transferred to one of the electrons in the P and the electron is lifted to the S band where it is free of wander around just as the conductance electrons flow around in metallic silver.

2. Another concept which was applied to theory of photolysis by Gurney and Mott was the presence of interstitial ions in a crystal lattice. A perfect crystal would contain some ions which are not in their proper places but in the interstitial positions. Presumably these are caused by thermal motion of the ions in the crystals. These ions and their movement through the crystal, either by going from the interstitial position to another or by moving to one of the holes which has been lifted by another interstitial ion, account for the every small electrical conductivity of crystals in the dark.

   These ions might be considered as being in solution in a crystal just a salt is dissolved in water. The conductivity of such crystals follows the same laws as the conductivity of ionised salt solutions. This conductivity decreases with decreasing temperature.

3. Gurney and Mott also assumed that silver bromide contains particles of colloidal silver.

4. In order to apply his theory to the formation of latent image as well as to the photolytic formation of metallic silver he postulated that the sensitivity specks, silver sulphide or whatever their complete composition might be, can serve the same function as the colloidal silver specks, *i.e.,* they can act as traps for electrons in the conductance level of the crystal. The theory involves the following points.

(i) When a film is exposed to light radiation, the primary absorption of light occurs in Br– ions of silver bromide crystal.

$$Br^- + hv \rightarrow Br^- + le$$

producing an excited electron.

(ii) The bromine atom is termed as a positive hole and is transferred from its original position in the crystal lattice to the surface of the crystal by the following process.

$$Br + Br^- \rightarrow Br^- \; Br$$

On the surface, the positive hole is trapped by coming in contact with an impurity centre line Ag_2S, probably by absorbing an electron from the sensitizer.

(iii) The excited electrons [from (i)] are free to move on through the crystal lattice. The moving electrons come in contact with a specks of silver and the trapped. This charges the speck negatively and this negative charge acts at some of interstitial silver ions which migrate to speck.

The positive charges on the speck due to silver ions are neutralised by the excited electrons.

$$Ag^{+} + le \rightarrow Ag.$$

Thus, the above process is repeated and a large number of silver atoms collect as an invisible speck at one point in a crystal. The same process is repeated in other silver bromide crystals on the film and ultimately leads to the formation of latent image.

Objections : These are.

(i) This theory had not taken sufficient account of the possibility of the recombination of photo released electrons with the positive holes which are formed at the same time.

$$\underset{\text{positive hole}}{Br} + le \rightarrow Br^{-}$$

(ii) *Gurney* assumed the existence of interstitial silver ions on a photographic film. *Mitchell* showed that a very small number of interstitial silver ions is found to exist on a photographic plate. This number is not sufficient to explain Gurney-Mott's theory.

IV. Mitchell's Theory (1957) : It involves the following steps.

(i) Upon the absorption of a quantum of light, a positive hole and photo-electron are formed.

$$Br^{-}\ hv \rightarrow \underset{\substack{\text{Positivee} \\ \text{hole}}}{Br} + \underset{\substack{\text{Photo} \\ \text{electron}}}{le}$$

(ii) The positive hole is trapped at a sensitivity speck in the crystal and this liberates an interstitial silver ion.

(iii) The liberated interstitial silver ions absorbed at some distortion or kink in the crystal lattice.

Then, the absorbed positively charged silver ion absorbs the photo-electron.

$$Ag^+ + 1e \rightarrow Ag.$$

This single silver atom is called the latent pre-image.

(iv) Another interstitial silver ion released by the trapping of a positive hole is absorbed this latent pre-image.

$$Ag^+ + Ag \rightarrow \begin{matrix} Ag^+ \\ | \\ Ag \end{matrix}$$

Again, the silver ion traps a photo-electron to give a two-atom specks. This is called latent sub-image.

$$\begin{matrix} Ag^+ \\ | \\ Ag \end{matrix} + 1e \rightarrow \begin{matrix} Ag \\ | \\ Ag \end{matrix}$$

(v) A third interstitial silver ion released by trapping of positive hole is absorbed on the latent sub-image and again, the absorbed silver ion traps a photo-electron, leading to the formation of latent image.

General Views : The series of experiments carried out in recent years have shown that the above theories are not satisfactory and they require further investigations.

SOME SOLVED PROBLEMS

Problem 1:

Find the recoil speed of hydrogen atom after it emits a photon in going from $n = 3$ to $n = 1$ state. Electron mass is 9.11×10^{-31} kg and $h = 6.626 \times 10^{-34}$ J s.

Solution:

The energy of the hydrogen atom in the n th state is given by

$$E_n = -\frac{Rhc}{n^2},$$

where R is Rydberg constant. From this, we get

$$E_1 - E_3 = -\frac{Rhc}{1^2} + \frac{Rhc}{3^2} = -\frac{8}{9}\ R\ h\ c.$$

The energy of the emitted photon is

$$\Delta E = E_1 \sim E_3 = \frac{8}{9} \text{ R h c.}$$

The momentum of the photon is

$$p = \frac{\Delta E}{c} = \frac{8}{9} \text{ R h.}$$

By conservation of momentum, the recoil momentum of the hydrogen atom will be equal (and opposite) to the momentum of the emitted photon. The recoil speed of the atom is

$$v = \frac{\text{momentum}}{\text{mass}} = \frac{8}{9}\frac{Rh}{m_H}.$$

But m_H = 1836 m, where m is electron mass.

$$\therefore \quad v = \frac{8}{9}\frac{Rh}{(1836\,m)}.$$

Putting the given values of h, m and using $R = 1.097 \times 10^7 \text{ m}^{-1}$, we get

$$v = \frac{8}{9}\frac{(1.097 \times 10^7 \text{ m}^{-1})(6.626 \times 10^{-34} \text{ Js})}{1836\,(9.11 \times 10^{-31} \text{ kg})}$$

$$= 3.86 \text{ m/s.}$$

Problem 2:

The ionisation potential of hydrogen tom is 13.6 volt. Find the wavelength of the Lyman series limit.

Solution:

When an atom absorbs energy so that its two stationary states becomes existed simultaneously, two superimposed sets of radiations are produced to give a 'beat' variation. If v¢ and v¢¢ represent the frequencies of two superimposed vibrations, the frequency v of the emitted beat is

$$v = v' - v''.$$

The wave mechanical vibrations are related to the energies of the corresponding state by the usual quantum expressions E′ = hv′ and E′ – E″ = hv″ so that

$$E' - E'' = h\,(v' - v'')$$

$$\Delta E = 13.61 \left(\frac{1}{(1)^2} - \frac{1}{\infty}\right) = 13.61 \text{ eV.}$$

Therefore, ionisations potential = 13.61 eV.

Problem 3:

A beam of monochromatic photons of energy 9 eV is incident on hydrogen gas all of whose atoms are in the ground state. It is found that the beam is fully transmitted without absorption. Why ? The ground state energy of an electron in the hydrogen atom is $E_I = -13.6$ *eV.*

Solution:

The minimum energy that can be absorbed by ground-state hydrogen atom is $E_1 \sim E_2$, which would excite it to the next state (n = 2). Now,

$$E_1 \sim E_2 = E_1 \sim \frac{E_1}{4} \qquad \left[\because E_n = \frac{E_1}{n^2}\right]$$

$$= \frac{3}{4} E_1$$

$$= \frac{3}{4} \times 13.6 = 10.2 \text{ eV.}$$

Hence photons of energy 9 eV cannot be absorbed by hydrogen atoms.

Problem 4:

Calculate the ionisation potential of hydrogen from the following date.

Solution:

According to Bohr's theory, the energy of an electron in the nth orbit of hydrogen atom is

$$E = -\frac{2\pi^2 e^4 m}{n^2 h^2}$$

When the atom is in the normal state, the only electron in the hydrogen atom in the first orbit, *i.e.*, n = 1.

$$\therefore \quad E = -\frac{2\pi^2 me^4}{h^2}$$

or

$$E = \frac{2 \times (3.14)^2 \times 9 \times 10^{-28} \times (4.8 \times 10)^4}{(6.6 \times 20^{27})^2}$$

$$= 2.165 \times 10^{-11} \text{ eg.}$$

To remove the electron from first orbit to infinity 2.165×10^{-11} erg of energy must be supplied. The amount of energy is called the ionisation potential of hydrogen atom.

Ionisation potential of hydrogen atom

$$= 2.165 \times 10^{-11} \text{ erg}$$

$$= \frac{2.165 \times 10^{-11}}{1.6 \times 10^{-12}} \text{ electron volts} = 13.53 \text{ eV.}$$

Problem 5:

Energy in a Bohr's orbit is given to be equal to $-B/n^2$ *where* B $= 2.179 \times 10^{-11}$ *erg. Calculate the frequency of radiation and also the wave number when the electron jumps from the third orbit to the second* ($h = 6.62 \times 10^{-27}$ *erg sec) orbit.*

Solution:

Energy of a Bohr's orbit is given by

$$E = \frac{B}{n^2} = \frac{2.179 \times 10^{-11}}{n^2} \text{ erg.}$$

For the third orbit, n = 3

$$E_2 = \frac{2.179 \times 10^{-11}}{9} = -0.2421 \times 10^{-11} \text{ erg.}$$

For the second orbit, n = 2

$$E_2 = \frac{2.179 \times 10^{-11}}{4} = -0.54475 \times 10^{-11} \text{ erg.}$$

$$E_3 - E_2 = (0.54475 \times 10^{-11}) - (0.2421 \times 10^{-11})$$

$$= 0.30265 \times 10^{-11} \text{ erg.}$$

When electron jumps from the third to the second orbit, a photon is emitted, the energy of which is given by

$hv = E_3 - E_2$ where v is the frequency

$$v = \frac{E_3 - E_2}{h} = \frac{0.30265 \times 10^{-11}}{6.62 \times 10^{-27}} = 4.572 \times 10^{14} \text{ sec}^{-1}.$$

The corresponding wave number is given by

$$\bar{v} = \frac{1}{\lambda} = \frac{v}{c} = \frac{4.572 \times 10^{14}}{3 \times 10^{-10}} = 15240 \text{ cm}^{-1}$$

Problem 6:

Give using spectral notation for the following states of the atom.

(i) n = 4, L = 2, S = 0

(ii) n = 4, L = 1, S = 1, J = 0 and

(iii) n = 3, L = 2, multiplicity 2.

Solution:

(i) Multiplicity (2S + 1) = 1, J = L + S = 2

∴ State will be 4 $^1D_2$.

(ii) Multiplicity (2S + 1) = 3 ∴ State will be 4 $^3P_0$.

(iii) Multiplicity will be (2S + 1) = 2 or S = 1/2

As L = 2, J = 5/2 or 3/2.

The two states which are positive are 3 $^2D_{5/2}$ or $^2D_{3/2}$. *Give using spectral notation for the following states of the atom.*

(i) n = 4, L = 2, S = 0

(ii) n = 4, L = 1, S = 1, J = 0 and

(iii) n = 3, L = 2, multiplicity 2.

Solution:

(i) Multiplicity (2S + 1) = 1, J = L + S = 2

∴ State will be 4 $^1D_2$.

(ii) Multiplicity (2S + 1) = 3 ∴ State will be 4 $^3P_0$.

(iii) Multiplicity will be (2S + 1) = 2 or S = 1/2

As L = 2, J = 5/2 or 3/2.

The two states which are positive are 3 $^2D_{5/2}$ or $^2D_{3/2}$.

Problem 7:

Calculate the energy in calories per mole or per Einstein for radiations of wavelength 100 Å..

Solution:

Energy per quantum = hv

But $h = 6.62 \times 10^{-27}$ erg-sec.

$$\therefore \quad V = \frac{c}{\lambda} = \frac{3\times10^{10}}{1000\,\text{Å}} = \frac{3\times10^{10}\,\text{cm.sec}^{-1}}{1000\times10^{-8}\,\text{cm}} = 3\times10^{15}\ \text{sec}^{-1}$$

Energy per quantum = hv

$$= (6.62 \times 10^{-27}\ \text{erg. sec})\ (3 \times 10^{15}\ \text{sec}^{-1})$$

$$= 19.86 \times 10^{-12}\ \text{ergs.}$$

But $N = 6.02 \times 10^{28}$ molecules/mole

and $hv = 19.86 \times 10^{-12}$ erg.

Therefore, the energy per mole or per Einstein

$$= N\,hv$$

$$= (6.02 \times 10^{23}\ \text{molecules mole}^{-1})\ 19.86 \times 10^{-12}\ \text{ergs/mole}$$

$$= 11.94 \times 10^{12}\ \text{ergs/mole}$$

$$= \frac{11.94\times10^{12}}{10^{7}}\ \text{Jule/mole} \qquad [\because\ 1\ \text{Joule} = 10^{7}\ \text{ergs}]$$

$$= \frac{11.94\times10^{12}}{4.184\times10^{7}}\ \text{cal/mole} \qquad [\because\ 1\ \text{cal} = 4.184\ \text{Joules}]$$

$$= \frac{2.8590\times10^{5}}{10^{3}}\ \text{K cal/mole} \ [\because\ 1\ \text{K cal} = 10^{3}\ \text{cal}]$$

$$= 285.90\ \text{K cal/mole.}$$

$$= \frac{285.90}{23.06}\ \text{electron volt} \ [\because\ 1\ \text{eV}=23.06\ \text{K cal/mole}]$$

$$= 12.390\ \text{electron-volts.}$$

Problem 8:

$$B \rightarrow C$$

1.0×10^{-5} mole of B was formed on absorption of 6.62×10^{7} ergs at 3600 Å. Calculate the quantum yield or efficiency.

Solution:

No. of moles reacting

$= 1.0 \times 10^{-5} \times 6.02 \times 10^{23}$ molecules

$= 6.02 \times 10^{18}$ molecules

No. of quanta absorbed

$$= \frac{\text{Total energy absorbed}}{\text{Energy of one quantum}} = \frac{6.62 \times 10^{7} \text{ ergs}}{h\nu}$$

$$= \frac{6.62 \times 10^{7}}{hc/\lambda}$$

$$= \frac{6.62 \times 10^{7} \times \lambda}{hc} \qquad \left[\because \nu = \frac{c}{\lambda}\right]$$

But $\quad l = 3600 \text{ Å} = 3600 \times 10^{-8}$ cm.

$c = 3 \times 10^{10}$ cm/sec, $h = 6.62 \times 10^{-27}$ erg/sec.

No. of quanta absorbed $= \frac{6.62 \times 10^{7} \times 3600 \times 10^{-8}}{6.62 \times 10^{-27} \times 3 \times 10^{10}} = 1.2 \times 10^{19}$

Therefore, equation yield $= \frac{\text{No. of molecules reacting}}{\text{NO. of quanta absorbed}}$

$$= \frac{6.02 \times 10^{18}}{1.2 \times 10^{19}} = 0.506.$$

Problem 9:

The bond energy in a molecule is 142.95 cal/mole. What is the longest wave-length of light capable of dissociating this molecule ?

Solution:

Energy per mole = N hν

where N= 6.02×10^{23} molecules, $h = 6.62 \times 10^{-27}$ erg-sec

$$\nu = \text{frequency} = \frac{c}{\lambda} = \frac{3.0 \times 10^{10}}{\lambda}$$

$\therefore$ Energy per mole $= \frac{6.02 \times 10^{23} \times 6.62 \times 10^{-27} \times 3 \times 10^{10}}{\lambda}$ ...(A)

The energy per mole = 142.95 cal/mole

$= 142.95 \times 10^{3}$ cal/mole

$= 142.95 \times 10^{3} \times 4.184 \times 10^{7}$ ergs/mole ...(B)

Equating eqs. (A) and (B), we get

$$\frac{6.02 \times 10^{23} \times 6.62 \times 10^{-27} \times 3 \times 10^{10}}{142.95 \times 10^{3} \times 4.184 \times 10^{7}}$$

$$= 142.95 \times 103 \times 4.184 \times 107$$

or $$\lambda = \frac{6.02 \times 10^{23} \times 6.62 \times 10^{-27} \times 3 \times 10^{10}}{142.95 \times 10^{3} \times 4.184 \times 10^{7}} \text{ cm}$$

$$= 2000 \times 10^{-8} \text{ cm} = 2000 \text{ Å}.$$

Problem 10:

The dissociation energy of hydrogen is 102900 cal/mole. If H_2 is dissociated by illumination with radiation of wave-length 2537 Å. What fraction of the radiant energy will be converted in to kinetic energy ?

Solution:

Dissociation energy = 102900 cal/mole

$$= \frac{102900}{6.023 \times 10^{23}} = 1.708 \times 10^{-19} \text{ cal/molecule.}$$

Also, one quantum of light is able to dissociate one molecule of hydrogen. Therefore, Energy of one quantum = $h\nu = \frac{hc}{\lambda}$.

But $h = 6.625 \times 10^{-27}$ erg sec,

$c = 3 \times 10^{10}$ cm/sec

$\lambda = 2537 \text{ Å} = 2537 \times 10^{-8}$ cm

$\therefore$ Energy of one quantum $$= \frac{6.625 \times 10^{-27} \times 3 \times 10^{10}}{2537 \times 10^{-8}} \text{ ergs}$$

$$= \frac{3 \times 6.625 \times 10^{-9}}{2537} \text{ ergs}$$

$$= \frac{3 \times 6.625 \times 10^{-9}}{2537 \times 4.18 \times 10^{7}} \text{ cal}$$

[$\because$ 1 cal 4.18 $\times$ 10^7 ergs]

But the dissociation energy per mole = 1.708×10^{-19} cal

$\therefore$ Energy converted into kinetic energy

$$= (1.874 \times 10^{-19} - 1.708 \times 10^{-19}) \text{ cal}$$

$$= 0.166 \times 10^{-19} \text{ cal}$$

Hence, fraction converted into K.E. = $\dfrac{0.166 \times 10^{-19}}{1.874 \times 10^{-19}}$.

Problem 11:

Calculate the time taken by the electron to traverse the first orbit in the hydrogen atom. Electron mass and charge are 9.1 × 10 31 kg and 1.6 × 10 19 C and h = 6.63 × 10 34 J-s.

Solution:

The radius of the n the Bohr orbit is

$r_n = 4\,\pi\,\varepsilon_0 \dfrac{n^2 h^2}{4\pi^2 m e^2}$ and the velocity of electron in the n th orbit is

$$v_n = \frac{1}{4\pi\varepsilon_0}\frac{2\pi e^2}{nh}.$$

The time taken by the electron to traverse the n the orbit is therefore

$$T_n = \frac{2\pi r_n}{v_n}$$

$$= 2\pi\,(4\,\pi\,\varepsilon_0)^2 \frac{n^2 h^2}{4\pi^2 m e^2}\frac{nh}{2\pi e^2}$$

$$= \frac{4\varepsilon_0^2 h^3 n^3}{me^4}.$$

For the first orbit, n = substituting the given values of h, m, e and ε_0 = 8.85 × 10^{-12} $C^2/N\text{-}m^2$, we get

$$T_1 = \frac{4.(8.85\times10^{-12})^2\,(6.63\times10^{-34})^3}{(9.1\times10^{-31})\,(1.6\times10^{-19})^4} = 1.5 \times 10^{-16}\ s.$$

Problem 12:

Find an expression for the radius of the electron orbit in the hydrogen atom in its n th state. What will be the approximate quantum number n for an electron in an orbit of radius 1 mm ? (Take required values from problem 1).

Solution:

The basic equation in the Bohr's theory of hydrogen atom are.

$$\frac{mv^2}{r} = \frac{1}{4\pi\varepsilon_0}\frac{e^2}{r^2} \qquad ...(1)$$

and

$$mvr = \frac{nh}{2\pi}. \qquad ...(2)$$

Squaring (2) and dividing by (1), we get

$$r = 4\,\pi\,\varepsilon_0\,\frac{n^2h^2}{4\pi^2me^2}.$$

Substituting the given values, we get

$$r = \frac{1}{9\times10^9\ Nm^2/C^2}\times \frac{n^2\,(6.63\times10^{-34}\ Js)^2}{4\times(3.14)^2\times(9.1\times10^{-31}\ kg)\times(1.6\times10^{-19}\ C)^2}$$

$$= 0.53\times10^{-10}\ n^2 \text{ meter}$$

$$= 0.53\ n^2\ Å.$$

For $r = 1$ mm $= 10^7$ Å, we have

$$10^7 = 0.53\ n^2.$$

$$\therefore \qquad n^2 = \frac{10^7}{0.53} = 18.87\times10^6 \text{ or } n;\ 4350$$

Problem 13:

The wavelength of the first line of Balmer series of hydrogen is 6562.8 Å. Calculate (i) the ionisation potential and the first excitation potential of the hydrogen atom. ($h = 6.63\times10^{-34}$ J-s,

$c = 3\times10^8 m\ s^{-1}$).

Solution:

(i) The wavelengths of the Balmer lines are given by

$$\frac{1}{\lambda} = R\left(\frac{1}{2^2}-\frac{1}{n^2}\right), \qquad n = 3, 4, 5, ...$$

For the first line n = 3 and l = 6562.8 Å = 6562.8×10^{-10} m.

$$\therefore \qquad \frac{1}{6562.8\times10^{-10}} = R\left(\frac{1}{2^2}-\frac{1}{3^2}\right) = \frac{5R}{36}$$

or $$R = \frac{36}{5 \times 6562.8 \times 10^{-10}\,\mathrm{m}} = 1.097 \times 10^{7}\ \mathrm{m}^{-1}.$$

This ionisation potential of an atom is numerically equal to the (ionisation) energy required to remove an electron completely from the atom in the normal state. In hydrogen atom, the electron stays in the first orbit (n = 1). Hence the energy required to remove this electron to infinity (where the energy is considered to be zero) is numerically equal to the energy of the electron in the first orbit. We know that electron energy in hydrogen atom is given by

$$E_n = -\frac{Rhc}{n^2}.$$

In the ground state, n = 1.

$$\therefore \quad E_1 = -Rhc$$

$$= -13.6\ \mathrm{eV}. \qquad \text{(as obtained in the leas problem).}$$

Here the ionisation potential of the hydrogen atom is 13.6 volts.

(ii) The first excitation potential of the atom is the energy required to shift the electron from n = 1 to n = 2 orbit, that is, $E_1 \sim E_2$. Now,

$$E_1 = -13.6\ \mathrm{eV}.$$

and $$E_2 = \frac{E_1}{4} = -3.4\ \mathrm{eV}. \qquad \left[\because E_n \propto \frac{1}{n^2}\right]$$

$$\therefore \quad E_1 \sim E_2 = 13.6 - 3.4 = 10.2\ \mathrm{eV}.$$

Hence the first excitation potential is 10.2 volts.

Problem 14:

The wavelength of a yellow line of sodium is 5896 Å. Calculate its wave number and frequency.

Solution:

The wave number $\bar{v}$ is the reciprocal of wave-length λ, that is,

$$\bar{v} = \frac{1}{\lambda}.$$

Here λ = 5896 Å = 5896 × 10^{-8} cm.

$$\therefore \quad \bar{v} = \frac{1}{5896 \times 10^{-8}\,\mathrm{cm}} = 16960 \text{ per cm.}$$

The frequency v is related to the wavelength l by

$$c = v\lambda,$$

where c is the speed of light. Thus

$$v = \frac{c}{\lambda} = \frac{3\times10^{8}\ \text{cm/s}}{5896\times10^{-8}\ \text{cm}} = 5.1 \times 10^{12}\ \text{s}^{-1}$$

Problem 15:

The average life-time of an electron in an excited state of hydrogen atom is about 10^{-8} s. How many revolutions does an electron in the n = 2 state make before dropping to the n=1 state? (R = 1.097 × 107 m^{-1}).

Solution:

Let v be the velocity of electron (mass m, charge e) in an orbit of radius r. This basic equations are.

$$\frac{mv^2}{r} = \frac{1}{4\pi\varepsilon_0}\frac{e^2}{r^2}$$

and $$mvr = n\frac{h}{2\pi}.$$

These equations give

$$v = \frac{nh}{2\pi mr},\quad r = \frac{n^2h^2\varepsilon_0}{\pi me^2}.$$

The number of revolutions of the electron in the orbit per second is

$$f = \frac{v}{2\pi r} = \frac{2Re}{n^3}$$

For n = 2 state, we have

$$f = \frac{2\times(1.097\times10^{7}\ \text{m}^{-1})\times(3\times10^{8}\ \text{ms}^{-1})}{8}$$

$$= 8.2 \times 10^{14}\ \text{s}^{-1}.$$

Hence the number of revolutions of the electron in its life-time of 10^{-8} second is $(8.2 \times 10^{14}) \times 10^{-8} = 8.2 \times 10^{6}$.

Problem 16:

An orange photon of wavelength 600 nm is emitted from an atom. Find the difference in energy in the two atomic states involved. Find the same result for the red 6563 Å line of hydrogen. (h = 6.63 × 10^{-34} j s, c = 3 × 10^{8} m s^{-1}, 1 eV = 1.6 × 10^{-19} J).

Solution:

By Bohr's postulate, the emitted frequency is given by

$$v = \frac{\Delta E}{h},$$

where ΔE is the difference in energy. But $v = c/\lambda$. Therefore

$$\frac{c}{\lambda} = \frac{\Delta E}{h} \text{ or } \Delta E = \frac{hc}{\lambda}.$$

Here $\lambda = 600 \text{ nm} = 600 \times 10^{-9}$ m.

$$\therefore \quad \Delta E = \frac{(6.63 \times 10^{-34}\,\text{Js}) \times (3.0 \times 10^{8}\,\text{m s}^{-1})}{600 \times 10^{-9}\,\text{m}}$$

$$= 3.31 \times 10^{-19}\ \text{J}$$

$$= \frac{3.31 \times 10^{-19}}{1.6 \times 10^{-19}} = 2.07 \text{ eV}.$$

For $\lambda = 6563$ Å $\times 10^{-10}$ m, we can show that

$$\Delta E = 3.03 \times 10^{-19}\ \text{J} = 1.9 \text{ eV}.$$

Problem 17:

What is the smallest wavelength in the spectral lines of Paschen series in the hydrogen atom ?

Solution:

The formula for Paschan series is

$$\frac{1}{\lambda} = R\left(\frac{1}{3^2} - \frac{1}{n^2}\right), \ n = 4, 5, 6, \ldots$$

For the smallest wave length, $n = \infty$.

$$\therefore \quad \lambda = \frac{9}{R} = \frac{9}{1.097 \times 10^{7}\ \text{m}^{-1}}$$

$$= 8.204 \times 10^{-7}\ \text{m} = 8204\ \text{Å}.$$

Problem 18:

The series limit of Balmer series is at 3646 Å. Calculate the wavelength of the first member Ha of this series.

Solution:

The formula for Balmer series is

$$\frac{1}{\lambda} = R\left(\frac{1}{2^2} - \frac{1}{n^2}\right), \ n = 3, 4, 5,$$

For series limit n = ∞ ad $\lambda = \lambda_\infty = 3646$ Å.

$$\therefore \qquad \frac{1}{3646 \text{ Å}} = \frac{R}{4} \text{ or } R = \frac{4}{3646 \text{ Å}}.$$

Again, for the first member n = 3.

$$\therefore \qquad \frac{1}{\lambda_1} = R\left(\frac{1}{2^2} - \frac{1}{3^2}\right) = \frac{5R}{36}$$

$$\text{or} \qquad \lambda_1 = \frac{36}{5R} = \frac{36}{5 \times \dfrac{4}{3646 \text{ Å}}} = 6563 \text{ Å}.$$

Problem 19:

The wavelength of the first line of Balmer series is 6563 Å. Calculate Rydberg constant.

Solution:

The wavelength of the spectral lines of Balmer series are given by

$$\frac{1}{\lambda} = R_H\left(\frac{1}{2^2} - \frac{1}{n^2}\right), \ n = 3, 4, 5, \ldots$$

For the first ($\lambda = 6563 \times 10^{-10}$ m), n = 3.

$$\therefore \qquad \frac{1}{6563 \times 10^{-10} \text{ m}} = R_H\left(\frac{1}{2^2} - \frac{1}{3^2}\right) = \frac{5}{36} = R_H$$

$$\text{or} \qquad R_H = \frac{36}{5 \times (6563 \times 10^{-10} \text{ m})} = 1.097 \times 10^7 \text{ m}^{-1}.$$

Problem 20:

Find the wavelength of the photon emitted when the hydrogen atom goes from n = 10 state to the ground state. ($R_H = 1.097 \times 10^{-3}$ Å^{-1}).

Solution:

Since the atom drops to the ground state (n = 1), the photon emitted belongs to the Lyman series. The wavelengths of the spectral lines in this series are given by

$$\frac{1}{\lambda} = R_H\left(\frac{1}{1^2} - \frac{1}{n^2}\right), \ n = 2, 3, 4, \ldots$$

For n = 10, we have

$$\frac{1}{\lambda} = R_H\left(1 - \frac{1}{100}\right) = \frac{99}{100} R_H$$

or

$$\lambda = \frac{100}{99\,R_H}$$

$$= \frac{100}{99 \times (1.097 \times 10^{-3}\ \text{Å}^{-1})} = 921\ \text{Å}.$$

Problem 21:

The first line of the Balmer series in the spectrum of hydrogen has a wavelength of 6563 Å. Calculate the wavelength of the first line of Lyman series in the same spectrum.

Solution:

The wavelength of the spectral lines of hydrogen spectrum are given by

$$\frac{1}{\lambda} = R\left(\frac{1}{n_f^2} - \frac{1}{n_i^2}\right),$$

where R is Rydberg constant. For the first member of Balmer series $n_f = 2$ and $n_i = 3$.

$$\therefore \quad \frac{1}{\lambda_1} = R\left(\frac{1}{2^2} - \frac{1}{3^2}\right) = \frac{5R}{36}.$$

For the first member of the Lyman series, $n_f = 1$ and $n_i = 2$.

$$\therefore \quad \frac{1}{\lambda_1'} = R\left(\frac{1}{1^2} - \frac{1}{2^2}\right) = \frac{3R}{4}.$$

From (i) and (ii), we have

$$\frac{\lambda_1}{\lambda_1'} = R\,\frac{5R}{36} \times \frac{4}{3R} = \frac{5}{27}.$$

$$\therefore \quad \lambda_1' = \frac{5}{27}\,\lambda_1 = \frac{5}{27} \times 6563 = 1215\ \text{Å}.$$

Problem 22:

The first member of Balmer series of hydrogen has a wavelength of 6563 Å. Calculate the wavelengths of the second and fourth members.

Solution:

The wavelengths of the spectral lines of Balmer series are given by

$$\frac{1}{\lambda} = R\left(\frac{1}{2^2} - \frac{1}{n^2}\right), \; n = 3, 4, 5, ...$$

For the first member, n = 3.

$$\therefore \quad \frac{1}{\lambda_1} = R\left(\frac{1}{2^2} - \frac{1}{3^2}\right) = \frac{5R}{36} \qquad ....(1)$$

For the second member, n = 4.

$$\therefore \quad \frac{1}{\lambda_2} = R\left(\frac{1}{2^2} - \frac{1}{4^2}\right) = \frac{3R}{16} \qquad ...(2)$$

Dividing eq. (2) by (1), we get

$$\frac{\lambda_1}{\lambda_2} = \frac{3R}{16} \times \frac{36}{5R} = \frac{27}{20}.$$

$$\lambda_2 = \lambda_1 \times \frac{20}{27}.$$

But $\lambda_1 = 6563$ Å (given).

$$\therefore \quad \lambda_2 = (6563 \text{ Å}) \times \frac{20}{27} = 4861 \text{ Å}.$$

Similarly $\lambda_4 = 4102$ Å.

Problem 23:

Calculate the (i) energy required to ionise hydrogen atom in the ground state (ii) limit of Balmer series. (Rydberg constant is 1.097×10^7 m^{-1}, $h = 6.63 \times 10^{-34}$ J s, c 3.0×10^8 m s^{-1}, 1 eV = 1.6×10^{-19} J).

Solution:

(i) The energy required to ionise, that is, to remove an electron from the hydrogen atom in the ground state (n = 1) to infinity (where the energy is zero) is numerically equal to the energy of the electron in the n = 1 orbit. Now, the energy of the electron in n th orbit of hydrogen is given by

$$E_n = -\frac{me^4}{8\varepsilon_0^2 h^2}\left(\frac{1}{n^2}\right)$$

$$= -\frac{Rhc}{n^2}. \qquad \left[\because R = \frac{me^4}{8\varepsilon_0^2 ch^3}\right]$$

For the ground (first) orbit, n = 1.

$$\therefore \quad E_1 = -\,Rhc$$

$$= -\,(1.097 \times 10^7\ m^{-1})\,(6.63 \times 10^{-34} J)\,(3 \times 10^8 m\ s^{-1})$$

$$= -\,21.8 \times 10^{-19}\ J$$

$$= -\,\frac{21.8 \times 10^{-19}}{1.6 \times 10^{-19}} = -\,13.6\ eV.$$

Hence the energy required to remove the electron from n = 1 orbit to infinity is 13.6 eV.

(ii) The Balmer series is represented by

$$\frac{1}{\lambda} = R\left(\frac{1}{2^2} - \frac{1}{n^2}\right),\ n = 3, 4, 5, \ldots$$

For the series limit n = ∞ so that

$$\frac{1}{\lambda} = \frac{R}{4}.$$

$$\therefore \quad \lambda = \frac{4}{R} = \frac{4}{1.097 \times 10^7\ m^{-1}}$$

$$= 3.646 \times 10^{-7}\ m = 3646\ Å.$$

Problem 24:

With Franck-Hertz type of experiment on sodium, the first spectral line to appear is the D-line, $\lambda = 5.89 \times 10^{-7}$ m. What is the first excitation potential of sodium ? Given : h = 6.63 × 10 34 J, s, c 3 × 108 m/s and 1 eV = 1.6 × 10 19 J.

Solution:

Let V volt be the excitation potential. Then the (excitation) energy imparted to the electron will be eV joule, where e coulomb is the charge on the electron. This energy is re-emitted as photon (radiation) when the electron returns to the normal state. If v be frequency of the emitted radiation, the photon energy will be hv. Thus

$$eV = hv \quad \text{or } V = \frac{hv}{e}.$$

If l be the wavelength of the emitted radiation, then v = c/λ.

$$\therefore \quad V = \frac{hv}{e\lambda}$$

$$= \frac{(6.63\times10^{-34}\,\text{J s})(3\times10^{8}\,\text{m s}^{-1})}{(1.6\times10^{-19}\,\text{C})(5.89\times10^{-7}\,\text{m})}$$

$$= 2.1\ (\text{J/C}) = 2.1\ \text{V}.$$

Problem 25:

The first two excitation potentials of atomic hydrogen in Franck-Hertz experiment are 10.2 and 12.09 volts. Draw an energy level diagram and show all possible transitions for emission and absorption along their wavelengths.

(h = 6.63 × 10 34 J s, c = 3.0 × 108 m/s, 1 eV = 1.6 × 10 19 J.)

Solution:

The energy level diagram, and the *three* possible transitions (a), (b) and (c) for emission. The frequency of radiation resulting from the transitions (a) is given by

$$v_a = \frac{E_2 - E_1}{h},$$

and the corresponding wavelength is

$$\lambda_a = \frac{hc}{E_2 - E_1}. \qquad [\because c = v\lambda]$$

Now, from the Fig., $E_2 - E_1 = 10.2$ eV $= 10.2 \times (1.6 \times 10^{-19})$ J. Therefore

$$\lambda_a = \frac{(6.63\times10^{-34}\,\text{J s})\times(3.0\times10^{8}\,\text{m s}^{-1})}{(10.2\times1.6\times10^{-19}\,\text{J})}$$

$$= 1.216 \times 10^{-7}\ \text{m} = 1216 \times 10^{-10}\ \text{m} = 1216\ \text{Å}.$$

Similarly, $$\lambda_b = \frac{hc}{(E_3 - E_1)}$$

$$= \frac{(6.63\times10^{-34})\times(3.0\times10^{8})}{(12.09\times1.6\times10^{-19})}$$

$$= 1.026 \times 10^{-7}\ \text{m} = 1026 \times 10^{-10}\ \text{m} = 1026\ \text{Å}.$$

Also, $$\lambda_c = \frac{hc}{(E_3 - E_2)}$$

$$= \frac{hc}{(E_3 - E_1) - (E_2 - E_1)}$$

$$= \frac{hc}{(12.09 - 10.2)\,eV} = \frac{hc}{1.89\,eV}$$

$$= \frac{(6.63 \times 10^{-34}) \times (3.0 \times 10^{8})}{(1.89 \times 1.6 \times 10^{-19})}$$

$$= 6.567 \times 10^{-7}\ m = 6567 \times 10^{-10}\ m = 6567\ Å.$$

In absorption, only the transitions starting from n = 1 shall be observed which correspond to 1216 Å and 1026 Å.

Problem 26:

A positronium atom is a system consisting of a position and an electron. Calculate the reduced mass, the Rydberg constant and the wavelength of the first Balmer line for positronium. (Give $m = 9.1 \times 10^{-31}$ kg, $R_H = 1.09737 \times 10^{-3}$ $Å^{-1}$ and $H_\alpha = 6563$ Å).

Solution:

The positron has the same mass m as the electron and has equal but positive charge. The reduced mass of the electron-positron atom is therefore

$$m = \frac{(m)(m)}{m + m} = \frac{1}{2}\ m = 4.55 \times 10^{-31}\ kg,$$

while the reduced mass of electron in hydrogen is very nearly m.

The Rydberg constant $\left(\frac{\mu e^4}{8\varepsilon_0^{\,2} c h^3}\right)$ for positronium is therefore half that for hydrogen (with infinitely heavy nucleus). Thus

$$R_P = 1/2\ R_H = 0.54868 \times 10^{-3}\ Å^{-1}.$$

The wavelength of first Balmer line (H_α) for hydrogen atom is given

$$\frac{1}{\lambda_H} = R_H\left(\frac{1}{2^2} - \frac{1}{3^2}\right),$$

while that for positronium atom is $\frac{1}{\lambda_p} = R_P\left(\frac{1}{2^2} - \frac{1}{3^2}\right)$.

Thus $$\frac{\lambda_P}{\lambda_H} = \frac{R_H}{R_P} = 2.$$

$$\therefore \quad \lambda_P = 2\lambda_H = 2 \times 6563 = 13126\ Å.$$

2

Radioactive Detectors

INTRODUCTION

Detectors are used to study the nature and energy of the paricles ejected from the nuclei during nuclear interactions. To produce nuclear transformations in lighter elements high energy paricles are required, which can be produced by particle accelerators. The three main groups of nuclear radiations are :

(i) charged particles like protons, alpha paricles and beta particles

(ii) uncharged particles like neutrons

(iii) electromagnetic radiations. Either ionisation techniques or excitation of atoms can be used as basic principles for detectors.

IONIZATION CHAMBER

The principle used here is that the passage of a charged particle through a gaseous medium causes ionization of the gas. By collecting the ions the intensity of the incident radiation can be measured. The ionization chamber consists of a gas filled container. It is closed at both ends and provided with the thin mica window at one end. The two electrodes of this chamber are the outer metal cylinder, connected to the negative of the d.c. power supply. Thus a potential difference is maintained between the container and the electrode. The radiation to be detected passes through the mica window. Upon entering the chamber the charged particle or photons ionise the gas inside the chamber. The number of ion pairs produced depends on the intensity of radiation. The electrons produced by the ionisation are drawn towards the collector rod at the centre. The positive ions are accelerated towards the container.

It n is the total number of ion-pairs produced inside the chamber then the total charge collected by the electrode will be q = 2ne. (ne-positive charges). If C is the capacity of the electrodes the e.m.f. generated in the chamber by the passage of the charged particle is given by

$$E = \frac{q}{C} = \frac{2ne}{C} \qquad ...(1)$$

This e.m.f appears across the rod and the container and produces a current i. If R is the resistance in the circuit the ionization current is given by

$$i = \frac{2ne}{CR} \qquad ...(2)$$

The ionization currents produced are very small. So the potential difference across R is fed to an amplifier and the voltage pulse can be amplified and registered.

Different types of particles of the same energy produce different ionization currents in travelling through the chamber at the same pressure.

So ionisation chambers are used to study different charged particles like electrons, protons, positrons and b-particles and gamma rays do not produce enough ionization to give detectable pulses hence ionization chambers cannot be used effectively to detect such radiation. Ionisation chamber can be filled with different gases depending upto the nations to be studied [*e.g.*, ae, CO, N_2, argon, etc]

Geiger-Muller Counter. Geiger and Muller designed this detecting device in 1928 which is still an efficient detector. This is nothing but a number of ionization chambers with a high field applied to one of the electrodes.

Due to this high voltages electrons produced by ionization are accelerated to such a high velocity that they themselves produce further ionisation by collision with the neutral molecules. So ion pairs are sharply multiplied [10^4 to 10^6 times]. This process is called gas amplification this device is highly sensitive. This is the principle of G.M. C is a metal chamber containing air or some other gas at a pressure of about 10 cms of Hg.

W is a fine tungsten wire stretched along the axis of the tube, and insulated from the tube by the elbonite plugs *E* and *E*. A high potential of about 1000 to 3000 volts is maintained between the wire and the chamber, the wire being kept at a positive potential. When the ionising particles enter the counter due to ionisation a few ions are produced. Due to the high P. D. these ions are multiplied by further collisions. The electrons thus produced move towards the central wire, which is equivalent to a small current impulse which flows through the resistance *R*. Thus the critical potential is lowered momentarily, and a sudden discharge through the resistance R is caused. This speedy development across *R* is amplified and then is allowed to operate a mechanical counter. In this way single particle can be registered. The sudden discharge clears the ions from the chamber and the counter is ready to register the arrival of the next particle.

A plot of the counting rate versus the counter potential. There is a threshold applied voltage below which the counter does not work. When the applied potential is increased the count rate rises rapidly to a flat portion of the curve called the plateau. In this region of operation, the counting rate is independent of small changes in the applied voltage when the applied electric field is very high, a continuous discharge takes place and the count rate increases very rapidly. In this region there is no necessity for any ionising event and so the tube must not be used in this region.

Merits

(i) It produces a large pulse which requires no further amplification.

(ii) The pulse size is independent of the nature of the incident radiation.

(iii) It is very sensitive because the radiation serves only to trigger a discharge.

Demerits

(i) It is insensitive for a period of 200 to 400 n sees following each pulse which prevents its use at very high counting rates.

(ii) It cannot provide information as to the nature of the particle which causes a pulse.

Wilson's Cloud Chamber. C.T.R. Wilson developed a cloud chamber which makes the tracks of charged particles to be seen. When a gas containing a vapour at saturation pressure is expanded rapidly it gets cooled so the gas becomes supersaturated.

If dust particles are present, condensation of vapour takes place and droplets are formed around the dust particles. In a dust-free atmosphere,

Wilson found that condensation could be produced on any negative ion present in the gas if the vapour could be expanded to more than 1.25 times of its initial volume. He showed that if the expansion ratios between 1.31 and 1.4 are used condensation takes place on the positive ions left by the passage of a charge particle through the chamber.

It consists of a cylindrical glass chamber C. It is provided with a glass plate cover and a perforated metal plate base. The metal plate is covered by a dark velvet cloth *V.* Water is sprinkled on this cloth to provide water vapour sufficient to saturate the air in the chamber C. The rubber diaphragm D forms an air-tight seal between the chamber *C* and space *S* below it. *A* is an evacuated container. It is kept closed by a piston *P.*

When the piston *P* is withdrawn the space s is connected to the container *A*. The pressure in *S* suddenly falls. This pulls down the diaphragm *D,* hence the air in the chamber *C* undergoes an adialiate expansion. After the expansion the container *A* is closed by the piston *P* and air is slowly admitted into the space S through a small valve *V.* The diaphram *R* goes back to its original position. The pressure in the chamber C also returns to the original value. The second perforated plate *B* is kept below the diaphragm *D.* Its position can be adjusted so that the extent to which *D* is pulled down can be altered. Thus any derived expansion ratio can be had.

After the expansion tracks are formed by water droplets, condensed on charged ions. The chamber is illuminated and the tracks are viewed through the top glass plate. These tracks can also be photographed from above. After photographing the tracks, the ions are removed by applying a potential difference of about 100V between a metal ring *k* just below the top glass plate and the base of the chamber. The electric field sweeps the ions from the chamber. Thus the chamber is made ready for next operation.

α-particles ionise copiously and so their tracks are thick. Electrons due to their smaller mass produce less ionisation and so their tracks are thin. X-rays and γ-rays cause ejection of electrons from atoms. These produce characteristic electron tracks in the cloud chamber.

Demerits

1. It is impractical to build cloud chambers large enough to show the entire path of an extremely high energy particle. To observe

the paths and the reaction induced by high energy particles bubble chamber or photographic emulsions are used.

2. The cloud chamber is less sensitive as the duration of the supersaturation is of the order of one second. Also it needs 20 to 30 seconds to come to equilibrium before another expansion can take place.

Bubble Chamber

Glaser invented the bubble chamber in 1952. If the liquid is heated under a high pressure to a temperature well above its normal boiling point, a sudden release of pressure will leave the liquid in a superheated state. If an ionising particle passes through in the liquid within a few milliseconds after the pressure is released, the ions left in the track of a particle act as condensation centres for the formation of vapour bubbles. The vapour bubbles grow at a rapid rate and attain a visible size in a time of the order of 10 to 100μs.

Thus in a bubble chamber, a vapour bubble forms in a superheated liquid, whereas in a cloud chamber, a liquid drop forms in a supersaturated vapour. Thus an ionising particle passing through the supertheated liquid leaves in its wake of a trail of bubbles which can be photographed. The liquid hydrogen bubble chamber operating at a temperature of 27 *K*. A box of thick glass walls is filled with liquid hydrogen and connected to the expansion pressure system. To maintain the chamber at constant temperature, it is surrounded by liquid nitrogen and liquid hydrogen shields. High energy particles are allowed to enter the changer from the side window N. A sudden release of pressure from the expansion valve is followed by light flash and camera takes the stereoscopic view of the chamber.

The incoming beam triggers the chamber. The charge of the tracks can be indentified by the direction of their curvature in the magnetic field applied over the bubble chamber. From the curvature and the length of the track, the momentum and energy of the particle can be found. The bubble chamber is used to study particle interaction and to detect very high energy particles.

Merits

1. The bubbles grow rapidly and as a result the tracks are not likely to get distorted due to convection currents in the liquid.

2. The density of a liquid is very large when compared to that of a gas of even high pressure. Hence the chances of collision of high energy particles with a molecule of liquid are very much greater. Consequently there is a greater chance of their track being recorded. So the chances of recording events like cosmic ray phenomena are improved when compared with cloud chambers.

Nuclear Emulsion

An ionising particle passing through a photographic emulsion produces a latent image of the track of the particle. This when developed gives a permanent record of the particle in the form of deposited silver in the negative. C. F. Power developed special types of emulsions much thicker than the usual ones. They contain some ten times the usual proportion of the Ag halide grains. Such plates after penetration by an incident ionizing radiation are turned into black on developing. Thus they bring a visual track of the ionizing radiation. Nuclear emulsion plates are sensitive to all types of charged particles. As the track lengths are very short they had to be observed under a high power microscope.

Merits

1. Very high energy particles can be studied using this technique. The tracks are short because of the high density of the emulsions.
2. They offer a permanent record of a nuclear event occurring in it.
3. They are very useful in the study of new fundamental particles and in space exploration and in the study of cosmic rays. They can be taken in baloons for recording primary cosmic ray phenomena.

Due to the stopping power of emulsions, many short lived particles like mesons, hadronic baryons can be brought to rest in the emulsion before they decay.

Scintillation Counters

When α-particles strike the fluorescent screens, they produce tiny flashes of light called scintillations. The energy of the particle is converted to light energy. The apparatus consists of a phosphor which produces a tiny flash of light when a charged particle is passed through it. The phosphor used are anthracene, napthalene, NaCl saturated with silver and silver iodide activated with thallium. The thickness of the phosphor should be such that the absorption is complete in the phosphor.

The flash falls on the photosensitive cathode of the photomultiplier tube. The photo-cathode emits electrons by absorbing the photons from the phosphor. The number of electrons emitted at the photocathode is proportional to the energy of the original incoming particle. These electrons are accelerated towards a positively charged electrode called dynode D_1. The surface of the dynode is coated with caesium-antimony alloy. These emit many secondary electrons for every incident radiation are multiplied by dynode D_1, These are again accelerated towards a second dynode D_2. The process continues. In a photomuliple tube ten to eleven dynodes which are kept at progressively higher potentials are used. Electronic counting devices are used to count the particles.

Merits

(i) *They are capable of detecting particles whose time of arrival are separated by even less than* 10^{-6}. *They can count faster than G.M. Counters.*

(ii) *The electric pulse generated is proportional to the energy of the individual incident panicles.*

Hence energies of the individual particles can he measured.

(iii) *They can operate in air or vacuum.*

Film Badges

A film badge consists of a photographic film encased in a plastic holder. When exposed to radiations, they darken the grains of silver in photographic film. The film is developed and viewed under a powerful microscope.

As α- or β-paricles pass through the film, they leave a track of black particles. These particles can be counted. In this way the type of radiation and its intensity can be known. However, γ-radiation darkens the photographic film uniformly. The amount of darkening tells the quantity of radiation.

A film badge is an important device to monitor the extent of exposure of persons working in the vicinity of radiation. The badge-film is developed periodically to see if any significant dose of radiation has been absorbed by the wearer.

INTRODUCTION OF NUCLEAR FUSION

When lighter nuclei moving at a high speed are fused together to form a heavy nucleus, the process is called nuclear fusion. In other words, it is the process of fusing light nuclei together to form a single heavy nucleus, for example, the fusion of hydrogen to form deuterium or helium, lithium and hydrogen to form helium. All these reactions take place in the presence of very high temperature at about 10^8K, so they are called *thermonuclear reaction.* All the reactions are exothermic and release huge energies. The energy come from conversion of mass to energy by the relation $E = mc^2$. The mass of the final single nucleus formed as a result of fusion is always less than that of the individual lighter nuclei. Some of the examples of such reactions along with the energies released by them are detailed below :

$$_1H^1 + {}_1H^2 \longrightarrow {}_2He^3 + \gamma + 5.5 \text{ MeV}$$

$$_1H^2 + {}_1H^2 \longrightarrow {}_2He^4 + \gamma + 3.2 \text{ MeV}$$

$$_1H^3 + {}_1H^1 \longrightarrow {}_2He^4 + \gamma + 19.7 \text{ MeV}$$

$$_1H^3 + {}_1H^2 \longrightarrow {}_2He^4 + {}_0n^1 + 17.6 \text{ MeV}$$

$$_3Li^6 + {}_1H^2 \longrightarrow {}_2He^4 + {}_2He^4 + 22.4 \text{ MeV}$$

$$_3Li^7 + {}_1H^1 \longrightarrow {}_2He^4 + {}_2He^4 + 17.3 \text{ MeV}$$

It may appear that it is probably easier to bring about fusion process than that of fission. But fusion reactions not only require highest kind of skill, but huge temperature conditions for their initiation and thus require suitable triggering devices.

Energy Released in Fusion : In nuclear fusion, the mass of the product nuclei is less than the sum of masses of reacting nuclei. The difference between the masses of nuclei of reactants and products is converted into energy according to the Einstein's mass-energy relation.

When two heavy hydrogen nuclei are fused together to form a nucleus of helium, the following reaction takes place :

$$_1H^2 + {}_1H^2 \longrightarrow {}_2He^4 + h\nu$$

The mass of each $_1H^2$ atom is 2.01471 a.m.u. while the mass of helium nucleus is 4.00388 a. m. u. The difference in mass 4.02942– 4.00388 = 0.02554 a.m.u. is converted into energy. The energy per fusion is 24 MeV. In this way a tremendous amount of energy is released in fusion. The energy released in fusion is as much as 1000 times greater than that in the nuclear bomb. No device for controlling the release of fusion energy during fusion has been invented.

Temperature Required for Fusion : The fusion process can not be carried out easily as it requires temperatures of the order of 10^8K. As such high temperatures are developed during the explosion of atom bomb, the latter can be utilised to initiate reaction in a nuclear fusion.

As fusion requires a high temperature to initiate itself, the atoms are fully ionised and these ions and free electrons are moving about very rapidly. Inspite of electrostatic forces of attraction between ions and free electrons, these move much more independently of each other than at ordinary temperatures. The mixture is electrically neutral and the whole state is called *'plasma state'*, a sort of second gaseous state.

Nuclear Fusion Reactions in Stars (Sun) : It has been estimated that sun is giving out energy equally in all possible directions at the rate of 3.7×10^{33} ergs/sec. Energy production in stars (sun is only a small star) occurs through its over-all hydrogen-to-helium reaction, $4\,_1H^1 \longrightarrow {}_2He^4$ which occurs in steps in two different ways (i) and (ii) described below :

(i) ***Carbon cycle*** : *In* 1926, **Eddington** *suggested the energy in sun is due to the nuclear fusion in which hydrogen is converted into helium.*

Eddington's work was supported by **Bethe** and **Weizsacker** (1938) who independently proposed that a carbon-nitrogen cycle is responsible for the production of solar energy.

The cycle is :

$$
\begin{aligned}
{}_6C^{12} + {}_1H^1 &\longrightarrow {}_7N^{13*} + \gamma + Q_1 \\
{}_7N^{13} &\longrightarrow {}_6C^{13} + {}_{+1}e^0 + {}_0v^0 \text{ (neutrino)} \\
{}_6C^{13} + {}_1H^1 &\longrightarrow {}_7N^{14} + \gamma + Q_2 \\
{}_7N^{14} + {}_1H^1 &\longrightarrow {}_8O^{15} + \gamma + Q_3 \\
{}_8O^{15} &\longrightarrow {}_7N^{15*} + {}_{+1}e^0 + {}_0v^0 \\
{}_7N^{15*} + {}_1H^1 &\longrightarrow {}_6C^{12} + {}_2He^4 \\
4\,{}_1H^1 &\longrightarrow {}_2He^4 + 2\,{}_{+1}e^0 + 2\,{}_0v^0 + (Q_1 + Q_2 + Q_3)
\end{aligned}
$$

The value of the factor, $Q_1 + Q_2 + Q_3$, comes out to be 24.7 MeV. The annihilation of positron supplies an extra 2 MeV of energy. Hence, the total energy released is in fact 26.7 MeV. Since $_6C^{12}$ serves as a kind of nuclear catalyst, the above series of nuclear reactions is called the *carbon cycle.*

The net result of this cycle is that four protons are converted into a helium nuclei with the energy release of $\approx$ 26 MeV. The carbon simply acts as a catalyst, entering only in the rate determining step and is later regenerated. The rate of the over-all reaction may be approximated as,

$$700\ \beta\ [H]\ [C] \left(\frac{T}{15 \times 10^6}\right)^{20} \text{ erg/gm/sec.}$$

where β is the density of the reacting matter in *g/cc.,* [H] is the weight concentration of hydrogen and [C] that of carbon and nitrogen both. Since the rate of reaction is very much dependent on the temperature, such cycle would be important only in the hotter regions within the Sun.

(ii) ***Proton-proton chain reaction*** : E. Saltpeter, in 1953, proposed a more direct sequence of nuclear reactions, which need a somewhat lower temperature than that required by the carbon-catalysed Bethe cycle. Here the rate-determining reaction is the relatively slow proton-proton collision, resulting in the production of deuterium.

$$_1H^1 + {}_1H^1 \longrightarrow {}_1H^2 + {}_{+1}e^0 + {}_0v^0$$

which undergoes the following, relatively fast reactions that compete the process.

$$_1H^2 + {}_1H^1 \longrightarrow {}_2He^3 + \gamma$$
$$_2He^2 + {}_2He^2 \longrightarrow {}_2He^4 + 2{}_1H^1$$

The net result, therefore, is the same, *i.e.,* the fusion of four protons into one helium nucleus.

The positrons emitted are annihilated by free electrons with the production of gamma rays.

The energy released in *p-p* chain is the same as in the C–N cycle (26.7 MeV for each helium nucleus).

According to the modern views, the temperature of the central region of the sun is 15,000,000°C and the carbon-cycle is the main reaction. In the large inner regions at somewhat lower temperatures, the proton-

proton reaction dominates. The chief source of energy in the stars large and hotter than sun is the carbon catalysed reaction, while in case of stars smaller and cooler than sun is the direct proton-proton reaction.

The details of the reaction sequences within the stelllar interiors have been known with the help of the experiments on the reactions of light nuclei to obtain the measurable rate in the laboratory; the experiments are to be usually conducted at extremely high kinetic energy, which are hardly available in stars. These experiments have proved that in the proton-proton chain, there may be some important side reactions. At $T > 1.3 \times 10^7$K the following sequence is more important.

$$_2He^3 + 2He^4 \longrightarrow {}_4Be^7 + \gamma$$

$$_4Be^7 \longrightarrow {}_3Li^7 + \gamma + {}_{+1}e^0$$

$$_3Li^7 + {}_1H^1 \longrightarrow 2{}_2He^4$$

Controlled Fusion. The belief of certain scientists that fusion process can also be controlled like that of fission, had led to expenditure of a great deal of scientific manpower and money. Because of the fact that temperatures of millions of degrees are involved, no material walls can be close to the region in which fusion is to take place.

This has led to the use of a gaseous discharge, called a *plasma,* held suspended in space by the magnetic lines of force of an electro-magnet. A plasma is a mass of ionised atoms, molecules or electrons and can be produced in many ways, stream of ionised atoms injected into the central region of a solenoid called *"magnetic bottle."*

By adjusting the strengths of current in the field coils, *X, Y* and Z, the shape of the magnetic field can be modified at will and the plasma narrowed or widened at will. It can also be transferred from one tube to the other. Plasma studies by various research laboratories have offered promising results, but still a lot is required to be done to successfully realise the power. The goal does not, however, seem to be so far.

Applications of Nuclear Fusion (Hydrogen Bomb). The phenomenon of nuclear fusion gives rise to the formation of hydrogen bomb.

Hydrogen bomb is based on the fusion of lighter nuclei (such as hydrogen) to form heavy nuclei (of helium). Energy released is so enormous that it is about 1000 times that of an atomic bomb. In a hydrogen bomb, a mixture of heavier isotopes of hydrogen, deuteron and tritium, is enclosed in a space surrounding an ordinary atomic bomb. The temperature produced by explosion of atomic bomb is very high.

$$_1H^2 + {_1H^2} \longrightarrow {_2He^4} + \gamma + \text{large amount of energy}$$

Unlike the fission process, it has been possible to bring about fusion under controlled conditions.

The first hydrogen bomb was exploded in November 1952 in Marshall Islands. In 1953 Russian exploded a thermonuclear bomb which had power of 1 million tons of T N T.

A diagrammatic representation of the hydrogen bomb. The function of fission bomb (atom bomb) here is to provide an atmosphere of the required high temperature. The hydrogen bomb is a horrible thing, as its energy released is unlimited. Once the fusion starts, the temperature is maintained by the process itself and it goes ahead until the fusible material.

Table 1 : Differences between nuclear fission and nuclear fusion.

	Nuclear Fission
1.	This process can be brought about only in the nuclei of heavy elements.
2.	In this case the heavy nucleus splits up into the lighter nuclei of comparable masses.
3.	This reaction can be initiated at ordinary temperatures.
4.	The energy liberated is about 200 Mev.
5.	This can be controlled and is being used for peaceful purposes.
6.	Percentage efficiency of the energy conversion is comparatively less.
7.	Products are radioactive and fission fragments are dangerous.
8.	Fast neutrons are released.
9.	Probability depends on the nuclear cross-section for slow neutrons.
10.	Fuel is in either solid or liquid state.
11.	Fuel can be stored for any length of time.
12.	Sum of the masses of the fission products is less than the mass of the fissile element.
13.	Z^2/A should be greater. Only heavy elements can have such values.
14.	The reaction can be made self sustained and chain reaction is possible.

Cobalt bomb : The cobalt bomb is a hydrogen bomb in a cobalt case. When disintegrating by the explosion, the cobalt becomes highly radioactive due to the formation of Co^{60} thus increasing the destructive effectiveness of the hydrogen bomb.

Nuclear Fusion

1. This process occurs in the nuclei of light elements.
2. In this case the lighter nuclei fuse to form a heavy nucleus.
3. This reaction initiates at 10^8 K.
4. In this process, the energy released is 24 Mev.
5. This cannot be controlled.
6. Percentage efficiency of the energy conversion is high.
7. Fusion products are non-radio active and are not dangerous.
8. Nature of the ejected particles depends upon the type of the of thermonuclear reaction.
9. Probability depends on the temperature and density of plasma.
10. Fuel is in plasma state.
11. Fuel cannot be stored as no container is available for such very high temperature.
12. Sum of the masses of the fusionable isotopes should be greater than the compound nucleus formed on fusion.
13. Light elements are best suitable for fusion.
14. To sustain reaction energy must be provided to fuel materials continuously, *i.e.* a high temperature is to be maintained.

THERMONUCLEAR REACTIONS

It is not easy to fuse two light nuclei into a single nucleus since electric repulsive force between the nuclei is to be overcome. Fusion reactions can take place only at very high temperatures of the order of 10 to 10 K. Only at this temperature, the nuclei can overcome the coulombian repulsion. So these reactions are called thermonuclear reactions.

A star is able to control thermonuclear fusion in its core because of its strong self gravity.

From A. S. Eddingtons's Stellar structure equations, the progressive changes of pressure from its centre outwards, the magnitude of pressure to density and fall of temperature from core to outwards etc., can be calculated. From these equations stable models of stars were arrived at.

NUCLEAR FISSION

In 1939, **Hahn** and **Strastsman** discovered that when uranium **nucleus** was bombarded with neutrons, it splits up into two radioactive nuclei, *i.e.*, barium (Z = 56) and krypton (Z= 36). *The new process of splitting of a heavy nuclei into nuclei of nearly comparable masses with release of energy was called fission by* **Friesh** and **Meitner** in 1939.

A nuclear fission is generally accompanied by the release of a small number of neutrons also. Immediately after the annoucement of the discovery of nuclear fission, physicists in many laboratories repeated and confirmed these experiments. Within next two years same process was confirmed for thorium and protactinium.

In addition to the two fragments, on the average 3 neutrons and about 200 Mev energy were released per fission. The energy was much larger than that released in any of the exoergic nuclear reactions.

Fission of other elements by projectiles other than neutrons: Detailed researches have shown that uranium is not the only element which undergoes fission. It has been found that elements having atomic numbers ≥ 85 undergo fission with suitable energy neutrons. In the case of U^{238} it succumbs to fission by fast neutrons whereas the U^{235} by slow neutrons or **what** are called thermal neutrons.

Further fission is also possible with other particles such as protons (6.9 MeV) deuterons (8.5 MeV), a-particles (32 MeV) and *y-rays.*

Main Points About Nuclear Fission

1. Fission Products. The natural uranium consists of three isotopes of atomic masses 238, 235 and 234 respectively. The abundance of U–235 is only 0.71% of the natural uranium which consists of 99.29% of U–238 and an extremely small percentage of U–234.

It was found that U–238 can be split up by very fast moving neutrons while U–235 can be fissioned by much slower neutrons. Moreover, the fission of U–235 is much more violent. When a slow neutron strikes the nucleus of ${}_{92}U^{235}$ it is captured and a highly unstable nucleus (${}_{92}U^{236}$), is produced which at once breaks up into two large parts with the emission of two or three neutrons. A large variety of products are possible.

By the carrier technique it was established that the products of fission of U–235 include isotopes of 35 elements having atomic numbers from 30 (zinc) to 64 (gadolinium). About 300 isotopes of these elements were found to be present. Most of these are β-radioactive with half-lives ranging from a small fraction of a second to many years.

It is accepted that only a few primary fission products are first formed which then give rise to secondary products.

Primary Processes

$$_{92}U235 + {_0}n^1 \longrightarrow {_{56}}Ba^{141} + {_{36}}Kr^{92} + 3\ {_0}n^1$$

$$_{92}U^{235} + {_0}n^1 \longrightarrow {_{54}}Xe^{139} + {_{38}}Sr^{95} + 2\ {_0}n^1$$

$$_{92}U^{235} + {_0}n^1 \longrightarrow {_{54}}Xe^{140} + {_{38}}Sr^{95} + {_0}n^1$$

Secondary processes.

$$_{54}Xe^{40} \xrightarrow{\beta^-} {_{55}}Cs^{140} \xrightarrow{\beta^-} {_{56}}Ba^{140} \xrightarrow{\beta^-} {_{57}}La^{140} \xrightarrow{\beta^-} {_{58}}Ce^{140}$$

The mass numbers of various isotopes obtained in nuclear fission extend over the range from 72 to 162. The yield of various fission products vary with mass numbers. From the graph it is evident that the fission of U–235 into two equal fragments is a very rare phenomenon. In fission, one fragment of comparatively larger mass and other of comparatively smaller mass are obtained.

2. Energy Released in Fission. In order to calculate the energy released in fission process, we will consider the following reaction ·

$$_{92}U^{235} + {_0}n^1 \longrightarrow {_{38}}Sr^{95} + {_{54}}Xe^{139} + 2\ {_0}n^1 + Q$$

Thus, mass equation is

$$235.10 + 1.009 \longrightarrow 93.945 + 138.955 + 2\ (1.009) + Q$$

When Q = 0.215 a.m.u.

$$= 9.215 \times 931 \text{ Mav} = 200 \text{ Mev approx.}$$

From the above calculations, it is evident that the energy released by fission of one uranium nucleus is approximately 200 MeV.

3. Release of Neutrons in Fission. Taliot and Van Halbon were first to show that average number of neutrons emitted per fission of uranium was 3.5 + 0.7. Fermi, however, got the average of 3 neutrons per fission of uranium. It is confirmed by now that whatever the mode of fission and whichever the fissioned material some neutrons are definitely ejected within an extremely short interval and are called *prompt neutrons.* The rest are emitted by fission products for an appreciable time fraction of a second to several seconds after the fission acts. They are called *delayed neutrons.*

4. Chain Reactions. When one nucleus of U^{235} undergoes fission by one neutrons, three neutrons are emitted. These emitted neutrons are emitted. These emitted neutrons can cause further fission of other U^{235} nuclei and this process continues. In this way a self-propagating of chain reaction is then possible liberating a tremendous amount of energy. The chain reaction. Self propagating of chain reaction did not become possibility till certain difficulties were overcome. The two chief causes that hinder self-propagation are :

(i) Leakage of neutrons from the system and

(ii) Presence of non-fissionable material in the system.

Let us discuss these one by one.

(i) ***Leakage of neutrons from the system :*** If a mass of the material to be fissioned is so small that a neutron incident on it cannot reach its nucleus but escapes outside then many of the neutrons of the initial reaction, shall be lost and the chain process will not be sustained. This leads to the idea of critical size or critical mass. If the system is such that the number of neutrons lost is more than the number produced, it is called sub-critical. If the number lost is just equal to the number produced, the system is said to be critical, and if the number of neutrons lost is much less than the number produced by the system, it is called over-critical. The chain reaction in such a case shall be self-sustained.

(ii) ***Presence of non-fissionable material in the system : This problem*** was overcome by carefully purifying the material and by neutralising the disturbing action of the impurities without actually removing them.

For the continuance of a chain reaction at least one out of the each 3 neutrons produced per fission must be preserved for further fission. If a start is made with an initial number of fast neutrons, some of these will be captured by a U–238 nucleus and if requisite activation energy is available, the fission would occur against each captured neutron. This is known as *fast fission factor,* (∈). The number of net effective fission neutrons produced per generation is sometimes known as *neutron multiplication factor* and is denoted as *K.* Depending upon the value of *K,* three situations might arise :

(i) If K < 1, the population of the neutrons decreases and the chain reaction will not be possible.

(ii) If K > 1, the population of the neutrons produced in the fission will increase and therefore the fission continues.

(iii) If K = 1, the rates of production and consumption of neutrons are same and such a process will not undergo explosion.

The neutron behaviour equation may be put as follows-

$$n = n_0 e^{\frac{K-1}{K},\ \frac{t}{\tau}} \qquad ...(1)$$

where, n_0 denotes the number of neutrons present at time $t = 0$ and n the number of neutrons at any time t. τ is the average life of a neutron.

If $$\frac{K-1}{K} = \rho \text{ (reactivity)}$$

$$n = n_0 e^{\frac{\rho t}{\tau}} \qquad ...(2)$$

This equation indicates the increase in number of neutrons with time.

For **K** = 1.1, we get $\rho \approx 0.1$ and then for a kilogram of uranium

$$n - n_0 e^{\frac{0.1t}{10^{-9}}}$$

if $$t = 10^{-6} \text{ sec.}$$

$$n = e^{\frac{1}{10^{-2}}} = n_0 e^{100}$$

$$\frac{n}{n_0} \approx 10^{43}$$

From this it follows that in a short interval of time neutrons are multiplied to such a great extent. These neutrons cause fission in about 10^{25} atoms. The energy released in just one micro-second from the fission of 1 kg uranium is 8.2×10^{13} Joules which results in a large explosion.

5. Nuclear Fuels : There are two classes of nuclear fuels :

(a) Fissile materials : There are the substances which can undergo fission directly with slow neutrons, liberating a large amount of energy. Examples are uranium–235, plutonium-239 uranium–233 etc.

(b) Fertile materials : A fertile material is one which though by itself non-fissible is converted into a fissile material by reaction with neutrons. For example. U–238 is converted into plutonium 239, a fertile material, by bombardment with neutrons (U–238 is not a fissile material).

6. Theory of Nuclear Fission : This has been explained by liquid drop model as nucleus of an atom has many similarities to a liquid drop. This model was proposed by Bohr and Wheeler. This model involves the following steps :

(i) *When a nucleus undergoes fission by a neutron, it combines with this incident neutron to form a compound nucleus which is highly energetic. For example, the uranium fission process could be represented as*

$$U^{235} + {}_0n^1 \longrightarrow [U^{236}]^*$$

where ${}_0n^1$ is a thermal neutron (0.025 Mev) and U^{236} is a compound nucleus. It is supposed that the compound nucleus, *i.e.*, $[U^{236}]^*$ retains its spherical shape by internal forces.

(ii) The additional energy added by the neutron initiates a series of rapid oscillations in the drop which at times assumes the shape of an ellipsoid (2.31B)

(iii) Now the restoring force of nucleus arises due to short range inter-nucleon forces. A condition is reached when oscillations become so violent that the ellipsoid will narrow (2.31C) down to a dumb-bell shape (2.31D). At this stage, dumb bell nucleus may be regarded to be corresponding to the state of boiling of the drop.

(iv) If the oscillations become very much violent, the system (2.31 D) will finally break at the neck into two major portions, due to the coulombic repulsion (2.31E).

It is important to remark here that there is a sort of threshold or critical energy required to produce a 'dumb bell' stage after which the nucleus cannot return to the initial spherical state because of the coulombian repulsion of two positive parts. The energy associated with stage D is called *critical energy* after which the nucleus cannot return to A and fission takes place.

The critical energy which must be supplied to the neutron, which is a potential energy diagram. In this diagram we see how the energy E_{crit} must be added to the system to enable the energy of the nucleus to become greater than the stability barrier energy E_b. Once the maximum barrier height has been overcome the system descends to the state of lowest potential energy and the fragments separate. When the mass of the compound nucleus is greater than the masses of the total fission fragments, fission is possible and the mass difference is released as energy. The value of the critical deformation energy E_{crit}, was first calculated by Bohr and Wheeler on the liquid drop model. They found.

$$E_{crit} = 0.89\ A^{2/3} - 0.02\frac{Z(Z-1)}{A^{1/3}}\ \text{Mev}$$

Application of Nuclear Fission. Three practical applications of the process of nuclear fission are: where A = Atomic mass of compound nucleus

Z = Atomic number

From this equation they concluded that even-even (even-proton even-neutron) nuclei require fast 1 neutrons and even-odd (even- proton odd-neutron) nuclei require slow neutrons to be fissionable.

(a) Atom Bomb, (b) Nuclear Reactor, and (c) Power Plant.

Let us discuss these one by one.

(a) Atom bomb : The first atom bomb used in Hiroshima utilized $_{92}U^{235}$ isotope as the main reacting substance and second bomb used in Nagasaki made use of $_{94}Pu^{239}$. The fission in both the cases is similar and uncontrolled. The mass of the substance used must exceed a certain quantity called the *critical mass.* It works on the fast neutron chain reaction, and brought about by bringing into contact rapidly two pieces of fissionable material such as U^{235} or Pu^{239}. The two pieces when separated are stable but the critical value and in consequence will explode very violently. The time period in which the different pieces are brought into contact is of the order of a millionth of a second. Enormous amount of energy equal to that produced by 20000 tons of TNT is produced accompanied by heat, light and radioactive radiations.

(b) Nuclear Reactor. It is a device to produce the nuclear energy in a controlled way to be used for peaceful purposes. It works on the same principle as the atomic bomb but under controlled conditions. All the neutron produced are not allowed to carry out the chain reactions.

In India, the first nuclear reactor was put into operation at Trombay in the year 1956 with the help of Canadian Government. In order to have an effective control on the fission process, cadmium rods are suitably inserted the fissionable material. Cadmium has the property of absorbing neutrons of all energies-cadmium rods are called arresters.

Types : Nuclear reactors of many kinds, shapes and sizes have been designed, both for research purposes and power generation. All these reactors can be classified into two groups depending mainly on the type of the moderator and the fissionable material used.

(i) ***Homogeneous Reactors*** : *In such reactors, heavy water* (D_2O) *is used as the moderator. The fissionable material, say uranium, can be taken in the form of solution of uranyl sulphate or the fuel can be suspended in* D_2O *in the form of very small particles.*

(ii) ***Non-Homogeneous Reactors*** : *In such reactors graphite is used as the moderator. The fissionable material is distributed through graphite in a regular manner forming a lattice.*

Numerous reactors of different designs have been constructed in all parts of world. However, all reactors possess some common features which are described as follows :

(i) ***Fuel*** : The material containing the fissile isotope is known as reactor fuel. The fuels which are commonly used are U^{235} in its natural concentration of 0.715% or in an enriched proportion, and Th^{233} and U^{238} which can be converted partly at least into fissile material.

(ii) ***Moderators and Reflectors*** : Materials used to reduce the neutron energy are known as moderators. Examples of moderators are graphite lime water, heavy water, beryllium and its oxide, and certain organic compounds.

The good moderators are usually materials of low mass number having small absorption cross section and large slowing down power (Sdp). The ordinary water has the greatest Sdp but it has an appreciable neutron absorption cross-section. However, it is used because it has low cost. The best moderator for neutrons is heavy water which has a very small cross-section and the highest moderating ratio of all moderators. Graphite has been employed as moderator or reflector in a number of reactors because it has a very small absorption cross-section and the second highest moderating ratio of all moderators.

Both beryllium and its oxide are regarded as good moderators because they have small absorption cross-section and the best Sdp of all the metals. However, the metallic beryllium is much more expensive than other moderators.

Hydrogenous organic compounds are used as moderators.

The above described materials are not suitable for use as the reflector for the reactor in which fast neutrons are used. For reflectors, the material of a high mass number must be used.

(iii) ***Reactor Coolant*** **:** The substance employed to remove heat that is being generated in the reactor core as a result of fission are known as reactor coolants. Examples of these are ordinary water, heavy water, liquid metals, organic liquids and gases. Each coolant has its own advantages and disadvantages.

(iv) ***Control materials*** **:** In order to achieve control in thermal reactors a neutron absorbing material is employed. The main characteristic is that it should not become radioactive as a result of neutron capture. Generally, cadmium in the form of rods is used. It can also be used at low temperature because it has low melting point.

In some reactors, an alloy of silver with 15% indium and 5% of cadmium is also used. This alloy has higher melting point and it has large nuclear cross-section for neutron capture.

Boron is also used in some reactors because it has very high melting point and large cross-section for neutron absorption, Boron is generally used in combination with stainless steel, aluminium or carbon.

In many reactors, the control elements are commonly located in the core in the form of either rods or plates. In some reactors it is put in the reflector close to the core.

(v) ***Reactor shielding*** **:** All nuclear reactors are hazard to persons in the immediate vicinity of the reactor. Therefore, the reactor core is covered by a radiation shield. It is generally a layer of concrete, about 2 to 3 m thick. This absorbs both γ-rays and neutrons.

In reactors operating at high powers, a shield of a few inches of iron or steel is used.

The schematic diagram of the first atomic pile at Chicago University. The reactor consists of small blocks of carbon, arranged together to form one solid mass of the size of a room. In between these blocks, lumps of pure uranium metal are inserted at regular intervals.

Long cylindrical holes are provided for the insertion or removal of fuel material, arrestors or detecting devices. The arrestors are cadmium rods which have the property of absorbing neutrons of the energies.

To operate such a pile, all but one of the cadmium rods are removed and the remaining one is slowly taken out, till the pile starts working.

Heavy water reactor has the advantage that it occupies less space as the neutrons are reduced to thermal energies after only 25 collisions, whereas for carbon, 115 collisions are required.

Nuclear reactors have been used :

(i) ***To produce* Pu^{239} *and other radioactive materials* :** For such purposes, the natural uranium which contains 99.3% of uranium–238 is kept in the nuclear reactor. When nuclear reactor is started, the uranium-238 present in natural uranium undergoes fission, liberating neutrons. These neutrons will convert uranium 238 into a long-lived plutonium-239 (half-life–24000 years).

(ii) ***To produce strong beam of neutrons* :** These neutrons may be used for making such isotopes that do not occur in nature. For this purpose, naturally occurring stable isotope is kept in a nuclear reactor and irradiation with neutrons converts it into a radioactive isotope. For example,

$$_{27}Co^{59} + {_0n^1} \longrightarrow {_{27}Co^{60}}$$

$$_{15}P^{31} + {_0n^1} \longrightarrow {_{15}P^{32}}$$

$$_{16}S^{32} + {_0n^1} \longrightarrow {_{15}P^{32}} + {_1H^1}$$

These isotopes are used in agriculture, industry and hospitals.

(iii) To generate power for driving the engine of ships, submarines, etc.

Nuclear Reactors in India : The following are the main reactors in India :

1. **Apsara Reactor :** It is the swimming pool reactor which consists of a lattice of enriched uranium fuel immersed in a large pool of water. The water in the pool plays the role of moderator, coolant and shield.

Apsara reactor was completed on August 14, 1956 at Trombay under the guidance of the late Dr. H. J. Bhaba. The chief advantages of this reactor are simplicity, low cost, safety, flexibility and accessibility.

2. **Cirus :** It is Second Indian research reactor which was completed on 10th July 1960. It was built in a joint Indo-Canadian project under the Colombo plan at Trombay. It is a natural uranium heavy water moderated and light water cooled high flux research reactor with a terminal power of 40 M. W. The standard fuel rod contains about 55 kg of natural uranium.

3. **Zerlina :** It is third Indian research reactor which attained criticality on Jan. 14, 1961. The name Zerlina stands for zero energy lattice investigation nuclear assembly as it was made to study the lattice assemblies of different kinds.

   The standard fuel rod contains about 45 kg of natural uranium. In this reactor, heavy water is used as a moderator.

4. **Purnima :** It is fourth India research reactor but it is first Indian experimental zero energy fast reactor. It attained criticality on May 22, 1972. It uses plutonium fuel as reactor.

5. **R-5 :** It is a thermal research reactor with a nominal power of 100 M. W. It is located adjacent to 40 M. W. Cirus reactor.

(c) Power Plant : One of the most important applications of nuclear fission is a nuclear power station.

Principle : Nuclear power station is based on the principle of conversion of nuclear energy into electrical energy. A huge amount of heat energy is released during nuclear fission. The heat produced is utilized in generating steam which runs the steam turbine. Electric generator is connected to the turbine, the latter acting as a prime mover.

Operation : Fission material such as U^{235} or Pu^{239} is subjected to nuclear fission in an atomic pile. An atomic furnace is an arrangement for controlling the chain reaction that starts once nuclear fission is done. If the chain reaction is not controlled, the result will be an explosion due to the fast increase in the energy released.

As a result of fission enormous heat is produced. The heat produced is given to the working substance (*i.e.,* sodium metal) which receives and carries out the heat to the heat exchanger.

In the heat exchanger, heat of the working substance is used to produce steam which runs the steam turbine. The turbine is connected to an electric generator. The electric power is obtained from the generator.

This type of power plant is set up at Tarapur, India.

Light-water Nuclear power plant. Most commercial power plants today are 'light-water reactors'. In this type of reactor, U–235 fuel rods are submerged in water. Here, water acts as coolant and moderator. The control rods of boron–10 are inserted or removed automatically from spaces in between the fuel rods.

The heat emitted by fission of U–235 in the fuel core is absorbed by the coolant. The heated coolant (water at 300°C) then goes to the exchanger. Here the coolant transfers heat to sea water which is converted into steam. The steam then turns the turbines, generating electricity. *A reactor once started can continue to function and supply power for generators.*

About 15 per cent of consumable electricity in U.S.A. today is provided by light water reactors. While such nuclear power plant will be a boon for our country, they could pose a serious danger to environment. In May 1986, the leakage of radioactive material from the Chernobyl nuclear plant in USSR played havoc with life and property around.

Disposal of reactor waste poses another hazard : The products of fission *e.g.*, Ba –139 and Kr–92, are themselves radioactive. They emit dangerous radiation for several hundred years.

The waste is packed in concrete barrels which are buried deep in the earth of dumped in the sea. But the fear is that any leakage and corrosion of the storage vessels may eventually contaminate the water supplies.

Breeder Reactor : We have seen that uranium-238 is used as a reactor fuel for producing electricity. But our limited supplies of uranium-238 are predicted to last only for another fifty years. However, non-fissionable uranium–238 is about 100 times more plentiful in nature. This is used as a source of energy in the so-called breeder reactors which can supply energy to the world for 5,000 years or more.

Here the uranium –235 core is covered with a layer or 'blanket' of uranium –238. The neutrons released by the core are absorbed by the blanket of uranium – 238. This is then converted to fissionable plutonium–239. It undergoes a chain reaction, producing more neutrons and energy.

$$^{238}_{92}U + ^{1}_{0}n \longrightarrow ^{239}_{94}Pu + 2^{0}_{-1}e$$

$$^{239}_{94}Br + ^{1}_{0}n \longrightarrow ^{90}_{38}Sr + ^{147}_{56}Br + ^{1}_{0}n$$

The above reaction sequence produces three neutrons and consumes only two. The excess neutron goes to convert more uranium to plutonium-239. *Thus, the reactor produces or 'breeds' its own fuel and hence its name.* Several breeder reactors are now functioning in Europe. However, there is opposition to these reactors because the plutonium so obtained can be used in the dreaded H–bomb.

Nuclear Power stations in India : These are as follows :

1. Tarapur Nuclear Power station : This became operational at 8.15 P.M. on April 1, 1969. This was constructed with the help of the Central Electric Company of U.S.A.

It is 420 M.W. energy station which consists of two enriched uranium reactors of the boiling water type. This station works well for some time but poses some difficulties which are being solved by Indian scientists.

2. Kota Nuclear Power Station : It is the second nuclear power station of 430 M.W. which is located near Kota at Rana Pratap Sagar in Rajasthan. It is heavy water moderated reactor using natural uranium as fuel. The first unit of the reactor became critical in 1972 and started feeding power in 1973 into the power system of Rajasthan.

3. Kalpakkam nuclear Power Station : It is 470 M. W. power station which has been constructed at Kalpakham about 100 km away from Madras. Its design is the same as that of Kota station.

4. Narora Nuclear Power Station : It is the fourth Indian atomic power plant which consists of two atomic reactors of 235 M. W. each and will deliver 440 M. W. of power to the Northern power grid when completed. It will use natural uranium as fuel and heavy water as moderator like the power station located at Kalpakkam and Rana Pratap Sagar.

Critical Mass

A uranium-235 nucleus on capturing a neutron undergoes fission. It splits up into two approximately equal daughter nuclei and ejects on the average 3 neutrons. These neutrons are emitted within about 10^{-15} second of the act of fission and are called *prompt neutrons.* They can be captured by other nuclei of U^{235} and bring about further fission, resulting in multiplication of neutron population and hence fission reactions at a terrific rate. The prompt neutrons formed as a result of first fission may reach only the boundaries of the mass of uranium and escape to the outside without suffering collision with a nucleus, if the mass of the fissible material is quite small. In such a situation, most of the fission neutrons would be lost to the outside instead of participating in a chain process which, therefore, would not be sustained. This leads to the concept of **critical size** or **critical mass.**

The continuation of a chain reaction requires that on an average, for every neutron lost by absorption or escape the surroundings at least one fresh neutron should be added by fission to the system. A system which just satisfies the above condition is called critical, whereas if there is system in which, for every neutron lost, the number of neutrons added is more than one, it is called over-critical, and build up chain reactions at a terrific rate; on the other hand, if the number of neutrons lost is more than that added, the system is called sub-critical.

It may be possible to arrange things such that a given mass of fisible material, which is sub-critical when in the form of a thin spherical shell, may become critical or overcritical when compressed in to a solid phere. This can be achieved by firing a suitable chemical explosive placed outside a hollow sphere. This is called *implosion principle or device.* This device was, used for exploding the first atomic bomb by Indian Scientists.

The critical size is comparable to the mean free path for the neutron. The latter has been found to be irreversibly proportional to the number of nuclei per unit volume and roughly speaking, the critical mass should be inversely proportional to square of density of the fissile material.

Dependence of Critical Mass on Temperature and Pressure

(i) If there is a rise in a temperature (during fission, etc), the density shall fall and the system which may be originally over-critical will become sub-critical and the chain reaction will stop.

(ii) If there is a rise in pressure, the density shall increase and the system which is critical in the beginning will become over-critical. This technique is utilised in modern atom bomb tests.

CHEMICAL EFFECTS OF RADIOACTIVE DECAY

In neutron capture reactions the recoil nucleus acquires high energy and causes perturbations in the orbital electrons. As a result, it breaks bonds with other atoms in the molecule and get separated. Such a recoil atom with high energy is called "hot atom", and the branch of chemistry involving the study of the excited recoil atoms is known as "hot-atom-chemistry". We shall be concerned in this portion with the hot atom reactions in both the organic and inorganic systems, and the formation of hot atoms from radioactive decay processes.

It is a fact that the recoil atom initially loses its high kinetic energy by elastic collision with the inactive isotope atom in accordance with the laws of conservation of momentum. In the course of collision, the recoil atom transfers a fraction of its energy to the colliding atom. However in a head-on collision, the recoil atom transfers a major fraction of its energy, as a result, its retention in the original compound increases.

We have stated earlier that the retention of the recoil atom is higher in the condensed (*i.e.*, solid or liquid) state than in the gaseous state. Such a high retention has been interpreted in terms of the "cage" theory of Libby according to which the recoil atom owing to the elastic collision becomes deprived of energy and remains imprisoned within the solvent "cage" of the condensed system.

On the other hand, the energy transfers in collisions with light atoms, *viz.*, hydrogen, carbon are very small, and the hot atom (*i.e.*, excited recoil atom) continues its collisions until it becomes "cooled" sufficiently for not able to collide further. Although, at this stage, it is no longer able to escape from the solvent cage, it still possesses kinetic energy that is sufficient to cause molecular dissociation.

In the latter stages of the dissipation of the recoil energy, a 'hot' atom may replace an isotopic atom or a group in other molecules of the target substance leading to a chemical reaction. Such chemical reactions are termed as "epithermal", and are independent of temperature.

For organic systems in hot-atom reactions, the retention of the I^{128} activity occurs in the 'hot' stage as in

$$CH_3 + I^* \longrightarrow CH_3{}^*I + I$$

But in the epithermal stage, the incompletely 'cooled' atom leads to the formation of a new molecule, thus,

$$CH_3I + I^* \longrightarrow CH_2II^* + H.$$

Similarly, the formation of labelled CH_2BrBr^* in the irradiation of $CH_2BrCOOH$, of labeled CH_3I^* and $C_2H_5I^*$ in the irradiation of iodine dissolved in CH_3CH_2OH, and of labeled $C_6H_5Br^*$ in the irradiation of aniline hydrobromide indicates that the excited halogen atoms can replace such groups as –COOH, –OH, $-CH_2OH$, $-NH_2$ and probably many others. The yield of the active atoms in one of these substitution products is usually less than about 10 per cent. Reactions of this type may be used to synthesize labeled compounds of high specific activity. The formation of C^{14}–labeled compounds in high specific activity, viz. labeled anthracene by slow neutron irradiation of nitrogeneous materials, such as acridine is known.

For inorganic systems involving hot-atom reactions, the retention of activity in oxygenated anions, such as permanganate, phosphate, and arsenate in thermal-neutron irradiations can be determined as functions of pH and concentration of the medium. The results have been interpreted by Libby on the basis of competition between hydration reactions and oxidative reactions for the hot atoms. For permanganate, the retention is independent of concentration in neutral and alkaline solution, and recoil causes the MnO_4^- ion to lose one or more O_2^- ions to give the manganese in the +7 state (*i.e.,*. MnO_3^+, MnO_2^{3+}, MnO^{5+} or Mn^{7+}). These unstable radical ions may then undergo either hydration reactions,

$$Mn{*}O_3^+ + 2OH^- \longrightarrow Mn{*}O_4^- + H_2O.$$

or, reduction by water,

$$4Mn{*}O_3^+ + 2H_2O \longrightarrow 4Mn{*}O_2 + 3O_2 + 4H^+$$

At $pH \geq 12$ the retention is nearly 100 per cent, because of the high probability of the hydration reactions. But with decrease of *pH,* the retention falls rapidly and reaches almost a constant value of 7% in the *pH* range 8 to 2 indicating that the reduction by water is faster than hydration in neutral and acid solutions. However, in highly acidic solutions the retention increases slightly perhaps due to the exchange reaction between the active and inactive manganese as in

$$Mn{*}O_3^+ + MnO_4^+ \longrightarrow MnO_3^+ + Mn{*}O_4^-$$

indicating that in acid solutions the retention increases with permanganate concentration. In arsenate irradiation, the is retention is nearly 100 percent over a wide *pH* range indicating that the hydration reactions strongly predominate. In phosphate bombardments the retention is independent of *pH* and is about 50 percent, presumably due to the fact that only about 50 per cent of the phosphorus in the primary recoil Fragment is in the + 5 state. Hot-atom reactions in a number of iodates, periodates, bromates,, and in different compounds of trivalent and hexavalent chromium, both as solids and in solutions have been reported.

Apart from nuclear reactions, hot atoms are also produced from the radioactive decay processes. The transformation of manganese into chromium in $MnO_4^- \longrightarrow CrO_4^{2-} + \beta^+$ or of tellurium into iodine in $TeO_3^2 \longrightarrow IO_3^-$ or of carbon into nitrogen in ethane in addition to the molecular disruption leads to a change n the structure of the compound. It should be stated here that the nucleus resulting from the β- decay must itself be radioactive. The 'hot-atom' chemistry has been especially studied in terms of the Li^6 ;(n, α) H^3, He^3 (*n,p*) H^3, Cl^{35} (*n, p*) S^{35}, Cl^{35}(n, α) P^{32}, S^{32} (*n, p*) P^{32} reactions, and of the active isotopes of carbon. In hot-atom reactions, however, difficulty arises to get quantitative information on the primary bond ruptures accompanying decay processes. But from a gas phase study of chemical effects of C^{14} decay n ethane, it has been shown that 50 per cent of the bonds, (*i.e.,* C–C bond) remain intact for giving C^{14} labeled methyl amine in decay of a C^{14} in a doubly labeled ethane molecule.

Finally, a separation of isomeric states may occur as a results of the recoil effect, especially when he isomeric transition takes place by the emission of conversion electrons. The γ-photon energies in isomeric transitions are much lower than the recoil energy in a neutron-capture process, often below 100 keV and rarely above 500 keV. This energy, however, seems insufficient to break a chemical bond, nevertheless, the missing electron must be replaced.

But the vacancy created in an inner electron shell by the internal conversion leads to the electronic rearrangements and emission of Auger electrons. The atom is, therefore, in a highly excited state (and positively charged), and if the atom is bound in a molecule, the molecular dissociation may take place. For the isomeric states of Br^{80} resulting from the (n, γ) reaction in bromobenzene, the 18-min Br^{80} can be separated from its parent 4.4/1 $Br^{80\,m}$ by the recoil effect analogous to the Szilard-Chalmers separation for bromine.

TRACERS IN CHEMISTRY

A radio-isotope can be detected by its radioactivity, and an inactive isotope of particular mass by means of mass spectrometer. Either of these properties may be utilised to trace the course of an element in various processes. An isotope used for this purpose is known as an isotopic tracer and the element which is labelled by the particular isotope is called a *tracer element.*

Table 2

Radioactive Isotopes	*Non- radioactive isotopes*
(i) Detection is very easy.	(i) Detection is very difficult,
(ii) They have existence for a very short time.	(ii) They survive or have existence for a very long time,
(iii) They are readily available in a large number.	(iii) They are not readily available,
(iv) Radiations emitted by these isotopes may prove harmful to the operator and to whom it has been injected.	(iv) They are not harmful at all.

It is important to remark that both types of isotopes are helpful in tracer technique. However, radioactive tracers are generally utilised more because of their easy detection.

Technique of Radioactive Tracer

The tracer technique is conducted as follows. First, a radioactive (stable) isotope is produced, then, from it, there is synthesized the required compound with a known specific activity sufficient for the prepared sample to be measured at the specified accuracy; after this is done the tracer experiment is started.

The amount of the isotope (in m curies) required for a study can be calculated from the formula :

$$q = A\,\frac{V_0}{V}\cdot\frac{100}{p}\cdot\frac{100}{\varphi}\cdot\frac{1}{3.7\cdot 10^7 \times 60}$$

where A = activity of sample which can be measured with specified accuracy

V_0 and V = volume or mass of substance under study and sample

p = fraction of radioisotope that finds its way into substance under study, per cent

φ = radiation recording coefficient, percent.

During or after the experiment a sample is taken and its radio-activity (isotopic composition) is determined. Radioactive radiation may alter the properties of solid compounds and affect a solvent, say, water and, hence, the substances dissolved in it. At high specific activities this effect must be taken into account.

Production of radioactive isotopes. The most important and efficient method of isotope production is by involving neutrons. Because of zero charge, the neutron is particularly effective as an entity for penetrating atoms to bring about nuclear reactions.

The simplest form of isotope production is neutron capture, that is entry of a neutron with simultaneous emission of y-radiation and known as (n, γ) nuclear reaction. For example, a radioactive isotope of sodium is produced by capture of a slow neutron.

$$_{11}Na^{23} + {}_0n^1 \longrightarrow {}_{11}Na^{24} + \gamma$$

The new atom produced is unstable and radioactive.

A much more useful method of isotope production is by (*n, p*) reaction. For example, a radioactive isotope of carbon is produced as follows :

$$_7N^{14} + {}_0n^1 \longrightarrow {}_6C^{14} + {}_1H^1$$

Similarly, radioactive sulphur and phosphorus are produced by the following scheme :

$$_{17}Cl^{35} + {}_0n^1 \longrightarrow {}_{16}S^{35} + {}_1H^1$$

$$_{16}S^{32} + {}_0n^1 \longrightarrow {}_{15}P^{32} + {}_1H^1$$

Some species, however, can be obtained only by bombardment with charged particles (*e.g.*, protons, α-particles in an accelerating machine such as a cyclotron). For irradiation in the pile, it is normal to use liquids or solids; for targets in the cyclotron solids are, of course, essential.

There are some isotopes which can be prepared by nuclear reactions of nuclei with charged particles and in such cases, in place of neutrons, fast deuterons are used. For example, radioactive sodium can be made by fast deuterons and magnesium.

$$_{12}Mg^{24} + {}_1H^2 \longrightarrow {}_{11}Na^{22} + {}_2He^4$$

However, the following are the main examples of radioactive isotopes:

C^{13} : A simple inexpensive method for producing enriched C^{13} as $BaCO_3$ has been developed recently, which makes use of the difference in the rate of the isotopic reactions. The dehydration of formic acid by sulphuric acid has been carried out for C^{13} enrichment.

Urey and coworkers have succeeded in the separation of C^{13} by exchange reaction method as given:

$$HC^{12}O_3^- + C^{13}O_2 \rightleftharpoons HC^{13}O_3^- + C^{12}O_2$$

Equilibrium is established by flowing a bicarbonate salt solution down a suitable column (counter current) to a stream of CO_2 gas. As shown by the equilibrium constant for the system, the bicarbonate solution is enriched in C^{13}.

C^{14} : One of the most important radioactive isotopes is radioactive carbon C^{14}. The atoms in the air are bombarded continuously by cosmic rays, and the nitrogen atoms in air are changed by the interaction of cosmic radiation into radio carbon C^{14}. It has a half-life of 5700 years, living plants absorb all forms of carbon through their intake of CO_2. But the radio carbon absorbed by the organisms during their life is not renewed but decays slowly after death.

So many methods are employed for enriching $C^{14}O_2$, one method involves in the pile neutron irradiation of NH_4Br, the target being dissolved in water. The radio carbon activities produced on dissolved pile neutron irradiated ammonium sulphate crystals in N–NaOH are subjected to fractionation. $C^{14}O_2$ is formed in large concentrations, along with other radio compounds. Separation of gases is made by differential adsorption and those of non-gaseous carrier may be carried out by distillation procedure, etc.

A standard method for determining carbon-14 in substances labelled with C^{14} has been recently developed. The labelled substance is first dissolved in a suitable solvent and its radioactivity is measured by proportional counter. This method is used for volumetric calculations. In solid samples some difficulties, such as deposition of film, combustion of sample, etc. arise and that is why solution form is desirable. The solvent must have the following characteristics :

(i) It should be able to dissolve a suitable quantity of the substance being analysed,

(ii) Its vapour should not construct a conducting path in that counter.

(iii) The vapour pressure of the solvent should be low enough so that the frequent concentration changes are not observed.

(iv) It should not ascend the sides of the container.

For this purpose ethylene glycol and dimethylformamide are most suitable. Carbon-14 has important applications in the synthesis of organic compounds. This isotope has following uses :

(i) In the study of the composition of organic compounds.

(ii) In the study of the formation of intermediate compound and its development.

(iii) In the formation of organic compounds labelled with carbon- 14.

(iv) In the formation of rates of absorption and emission of medicines.

(v) In the study of biosynthesis of carbohydrates, protein and aminoacids in the body.

Tritium ($_1H^3$ or T)

The name tritium is derived from the Greek word 'tritos' (meaning third). This isotope occurs in very small quantities in nature. Out of 10^{10} parts of ordinary water only 7 parts are tritium. It is also found in heavy water. When the volume of 75 tons of water, after being electrolysed, reduces to mere 0.5 c. c., it is shown by mass spectrograph that in 104 parts of it only 1 part of tritium is present.

Oliphant, Rutherford, Harteck and Dee obtained tritium by bombarding ND_4Cl, $(ND_4)_2SC_4$ D_3PO_4 etc., with deuterons.

$$_1D^2 + {_1D^2} \longrightarrow {_1T^3} + {_1H^1}$$

The above conversion occurs via the formation of helium nuclei which are very unstable due to its greater energy.

$$_1D^2 + {_1D^2} \longrightarrow 2{*}He^4 \longrightarrow {_1H^1} + {_1T^3}$$

The energy associated with the above helium nuclei is 2.3 MeV more than that associated with ordinary helium nuclei making it unstable.

The tritium is also obtained by bombarding lithium by slow moving neutrons.

$$_3Li^6 + {_0n^1} \longrightarrow {_1H^3} + {_2He^4}$$

Tritium may also be obtained by the bombardment of neutrons on beryllium and of neutrons on boron and nitrogen.

$$_4Be + {_1D^2} \longrightarrow {_4Be^8} + {_1H^3}$$

$$_5B^{10} + {_0n^1} \longrightarrow {*_5B^{11}} \longrightarrow \underset{\substack{\downarrow \\ 2_2He^4}}{_4Be^8} + {_1H^3}$$

$$_5B^{11} + {_0n^1} \longrightarrow {_5{*}B^{12}} \longrightarrow {_4Be^9} + {_1H^3}$$

$$\longrightarrow {_6C^{12}} + {_1H^3}$$

$${}_{7}N^{14} + {}_{0}n^{1} \longrightarrow {}_{7}{}^{*}N^{15} \longrightarrow 3\,{}_{2}He^{4} + {}_{1}H^{3}$$

$${}_{7}N^{15} + {}_{0}n^{1} \longrightarrow {}_{7}{}^{*}N^{16} \longrightarrow {}_{6}C^{13} + {}_{1}H^{3}$$

Tritium is a β-radioactive isotope of hydrogen and converts into a lighter isotope of helium after emitting β-ray.

$${}_{1}T^{3} \longrightarrow {}_{2}He^{3} + {}_{-1}e^{0}$$

Tritium is used as tracer in solvent extraction. The age of wine and many other products is determined using tritium. Libby (1953) determined the amount of tritium in rain water and predicted the time for which moisture has stayed in air between evaporation from sea and condensation. The amount of tritium also helps in predicting the time for which water has stayed in water reservoir underground.

P^{30} : It is made by bombarding P_{31} with high energy γ-rays. The γ-rays of energy 17 Mev

$${}_{3}Li^{7} + {}_{1}H^{1} \longrightarrow {}_{4}Be^{8} + \gamma$$

are derived by the above reaction.

In the commonest type of reaction a neutron is evolved from the target.

$${}_{15}P^{31} + \gamma \longrightarrow {}_{15}P^{30} + {}_{0}n^{1}$$

Other methods involve the following types of nuclear reactions :

$${}_{16}S^{32} + {}_{1}D^{2} \longrightarrow {}_{15}P^{30} + {}_{2}He^{4}$$

$${}_{14}Si^{30} + {}_{1}H^{1} \longrightarrow {}_{15}P^{30} + {}_{0}n^{1}$$

P^{32}. The following type of nuclear reaction, is employed for preparing P^{32}.

$${}_{15}P^{3}1 + {}_{1}H^{2} \longrightarrow {}_{15}P^{32} + {}_{1}H^{1}$$

In this method red phosphorus was bombarded with 2 Mev deuterons.

${}_{15}P^{32}$ decays by β-particle emission.

$${}_{15}P^{32} \longrightarrow {}_{16}S^{32} + {}_{-1}e^{0}$$

Other nuclear reactions yielding P^{32} are as follows :

$${}_{17}Cl^{35} + {}_{0}n^{1} \longrightarrow {}_{15}P^{32} + {}_{2}He^{4}$$

$$_{16}S^{32} + _{0}n^{1} \longrightarrow _{15}P^{32} + _{2}H^{1}$$

$$_{14}Si^{29} + _{2}He^{4} \longrightarrow _{15}P^{32} + _{1}H^{1}$$

I-131 : It is the pioner radio iodine tracer for the study and treatment of the thyroid function in human being, I-131 is produced as fission product (the mass 131, chain yield for thermal neutron fission is ~29%) or by the neutron irradiation of tellurium-130. The latter produces tellurium-131 which subsequently decays by beta particle emission to yield carrier free I-131.

$$_{52}Te^{130} + _{0}n^{1} \longrightarrow _{52}T^{131} + \gamma$$

$$\downarrow$$

$$_{53}I^{131} + b$$

The target is oxidised with dichromate and acid, when tellurate and iodate are formed; mild reduction with oxalic acid converts the iodate into free iodine, which is distilled off, trapped in dilute alkali, and reduced to iodide by sulphite. :

Separation of Radioactive Isotopes

1. Solvent Extraction. Solvent extraction method is rapid, and relatively specific and requires no commercial equipment. In the simplest case of solvent extraction the radioactive species is much more soluble in the organic solvent than in aqueous phase. For example, iodine in carbon tetrachloride, can be separated from others.

For the extraction of an ion, the first step is the formation of neutral species which is soluble in organic solvents. This may be done by complexing agents such as Cl^-, Br^-, O^{-2}, phcnan-throline, cupferon etc. If the complex is uncharged, it is ready for extraction; if charged, it is made neutral with convenient ions. Some solvents such as ethers and alcohols act as complexing agents themselves.

2. Chromatography : Various types of chromatographic procedures give clean separation in short time. The scope of these methods has been increased by the availability of large number of new solvents. Thin layer chromatography (TLC), paper chromatography, and ion exchange chromatography are used for the separation of radioactive isotopes.

3. Precipitation : Precipitation has been the old "reliable" method for many specific nuclides. In this method of separation, the radioactive isotope is first converted into solution and then precipitated by suitable reagents.

4. Electrochemical methods : The use of electrodeposition in separating the mixture of radio-isotopes has been limited. Controlled cathode deposition, for example, would be a valuable tool but less important than solvent exctraction method.

5. Physical methods : Distillation and volatilization methods may be used for radioactive isotopic separation. Radioactive tracer distillations are similar to conventional macro-level distillations. Such volatile substances as $TeCl_4$, SbH_3 or OsO_4 provide the basis for separation in a single step from a large number of contaminants.

Radio Isotopes as Tracers

When we use radio-isotope as a tracer, we have to keep the following points in mind :

(i) The choice of half-life of radio-active isotope :

When we use short lived isotopes, time is limited for observation. In the case of long lived isotope, the intensity of radiation emitted is very less and, therefore, cannot be easily detected. Thus, isotopes useful for tracer must have half-lives which are neither too short nor too long. Generally, isotopes having half-lives roughly between a few hours to few thousand years are used as tracers.

(ii) Detection of radiations :

Radiations from radioactive elements can be detected and measured in many ways. Following are some methods which are generally applied:

(a) Photographic measurement method.

(b) Electroscope method.

(c) lonisation chamber method.

(d) Geiger Muller or G.M. method.

(e) Scintillation counter method.

Any of the above methods may be used. However, the best method to employ in any case depends upon the type of radiation and energy of radiation. The last three methods are of great importance in measuring radioactivity. However, scintillation counter is a latest and most accurate method. It is replacing G.M. counters because of its several advantages.

NON-RADIOACTIVE ISOTOPES AS TRACERS

In experiments using non-radioactive isotopes as tracers, the apparatus used for their detection is the mass spectrometer. However, this method is difficult and requires certain precautions to be followed while determining and detecting their amounts by mass spectrometer.

RADIOACTIVE ISOTOPES

In 1913, *G. Hevesy* and *F. Paneth* suggested the use of radioactive elements as indicators or tracers and they determined the solubility of sparingly soluble salts.

Some elements occur in several isotopic forms and they, can be used as tracers as such. If isotopes do not occur in nature, they may be produced artificially by various methods.

These radioactive isotopes emit characteristic radiations or particles which will serve to prove the presence of these isotopes. This forms the basis of several analytical procedures:

1 Adsorption and Occlusion studies : A small amount of radioactive isotope is mixed with the inactive substance and the activity is studied before and after adsorption. Fall in activity gives the amount of the substance adsorbed.

Similarly .from the activity of a material adsorbed from a mixture of stable and radioactive isotopes, it is possible to estimate the extent of adsorption, the number of adsorbed molecules and the area covered by these.

2. Surface area of crystals : The principle involved in the determination of surface area of crystals is that the exchange of active and inactive isotopes occurs only at the surface and not inside the crystal lattice.

The loss of activity gives the surface area, *i.e.,* in the study of $PbSO_4$ a known amount of this mixed with solution of radioactive $PbSO_4$ until equilibrium is reached between molecule on the surface of the solid and in solution. Activity of the solid and solution is determined separately. Then,

$$\frac{\text{Pb in surface}}{\text{Pb in solution}} = \frac{\text{Radioactivity of solid}}{\text{Radioactivity of solution}}$$

3. Solubility of sparingly soluble salts : The solubility of lead sulphate in water may be estimated by mixing a known amount of radioactive lead with ordinary lead and dissolving the mixture of two in nitric acid followed by precipitation as $PbSO_4$ by adding H_2SO_4.

$$Pb^* (NO_3)_2 + 4HO_3 \rightarrow Pb^* (NO_3)_2 + Pb (NO_3)_2$$

$$Pb^* (NO_3)_2 + Pb(NO_3)_2 + 2H_2SO_4 \longrightarrow \frac{[Pb^*SO_4 + PbSO_4]}{ppt.} + 4HNO_3$$

The precipitated lead sulphate is filtered and the amount of lead sulphate still in solution as dissolved is determined by measuring the activity of radio lead (Pb) in solution. Therefore, from the radioactivity of the solution, the amount of $PbSO_4$ and hence that of $PbSO_4$ can be estimated.

4. Study of efficiency of analytical separations : The efficiency of analytical procedure may be measured by adding a known amount of radio-isotope to the sample before analysis begins. After the final determination of the element in sample, the activity of the precipitate is determined and compared with the activity at the start.

5. To follow the progress and completeness of chemical separations. Radioactive tracers have been found to be of great importance in following the progress and completeness of various chemical separations.

6. Ion exchange technique : Ion-exchange process of separation is readily followed by measuring the activity of successive fractions eluted from the column. This technique has been found to be of great use in establishing the order of elution of the rare-earth cations and in studying the separation of transuranic elements such as americium and curium from rare earth fission products and from one another.

ACTIVATION ANALYSIS

Introduction : It is one of the most sensitive and specific methods available for the determination of trace quantities of a wide range of elements. This technique was first introduced by G. Hevesy and H. Levi in Denmark in 1936. However the first systematic presentation of activation analysis as a method was made by Clark and Overman in 1947. As reactors and accelerators have become more common these days, this technique has therefore become as a powerful tool in many aspects of science and industry. Activation analysis can be performed by two methods:

Absolute Method

Theory : The essential basis of the method is that the element to be determined is made radioactive usually by slow neutron irradiation and then this radioactivity, after suitable chemical irradiation and if necessary, is a measure of the mass of the element originally present.

When an element is made radioactive by placing it in a homogeneous flux of energetic charged particles of neutrons, the activity 'A' produced in the element is given by :

$$A = N\phi\sigma\left(1 - e^{\frac{-0.693}{T_{1/2}}t}\right) \qquad ...(1)$$

where N = the number of atoms of the nuclide in the sample capable of forming the radio-isotope in question.

ϕ = the flux of bombarding particles in unit of particles per square per second,

σ = the isotopic cross section for the nuclear reaction in units of square centimeters per target atom (cm^2)

t = the time of irradiation.

$T_{1/2}$ = the half life of the nuclear species produced.

t and $T_{1/2}$ are expressed in the same units. The weight *w* of the element activated by bombarding particles is given by

$$w = \frac{N \times M}{6.023 \times 10^{23}}$$

or

$$N = \frac{w \times 6.023 \times 10^{23}}{M} \qquad ...(2)$$

where *M* is the chemical atomic weight of the element to be estimated Substituting equation (2) in (1) wc get

$$A = \frac{w \times 6.023 \times 10^{23}}{M}\phi\sigma\left(1 - e^{\frac{-0.693}{T_{1/2}} \times t}\right)$$

$$w = \frac{MA}{6.023 \times 10^{23} \times \phi\sigma\left(1 - e^{\frac{-0.693}{T_{1/2}} \times t}\right)}$$

If one knows the values of M, A, σ, ϕ, $T_{1/2}$ and t, the weight w (the amount of the sample to be determined) can be calculated by using equation (3).

Procedure for absolute method : The procedure for absolute method may include the following steps :

(i) *The first step of activation analysis is the activation of a sample to be analysed, i.e., the conversion of stable isotope of the sample to be analysed into radioactive nuclides. The activation is done by placing the sample in a flux of bombarding particles (charged or uncharged) or rays* (γ *or X rays*) *for a time long enough to produce the radio isotope of the element to be determined.*

The use of charged particle bombardments has been superseded by neutron activation because of the availability of a high constant neutron flux in the chain reacting piles. Therefore, this technique is generally termed as the neutron activation analysis.

(ii) After irradiation, dissolve the sample and add appropriate carriers. Then carry out chemical separations to isolate the element or a suitable compound free from all other radio-nuclides, *i.e.*, until it is radio chemically pure. The carrier and the element in the sample must be in the same chemical form.

(iii) Now determine the activity of the radio-chemical compound by the aid of a semiconductor detector.

(iv) Then, use the equation (3) to determine the amount of sample to be estimated.

Sensitivity of the absolute method : It is important to mention here that σ is not known with high accuracy; ϕ the flux of particles, is very difficult to measure exactly; and A, the activity, must be corrected for chemical yield, counting efficiency and counting geometry, Therefore, the weight w can not be determined with greater accuracy *i.e.,* the method is not very sensitive and it requires many modifications.

B. Comparator Method. The difficulties of the absolute method are avoided by using a comparator procedure, *i.e.*, simultaneous irradiation of sample and standard containing a known proportion of the element being determined.

In this method, analysis is carried out by bombarding a standard sample and the unknown sample under the same conditions. The chemical treatment of the two is earned out in the identical conditions. Carrier is used in the separation and isolation of the element to be analysed. Since both standard and unknown have been treated in an identical way, the weight w of the unknown the constituent to be determined, is obtained as follows :

$$\frac{\text{Wt. of X in unknown}}{\text{Wt. of X in standard}} = \frac{\text{Total activity from element X in unknown}}{\text{Total activity from element X in standard}}$$

Another modification of the comparator method: Irradiate the sample to be analysed for a known time. It is generally preferred that the radioactive species formed should produce gamma rays either by the direct emission or by annihilation of positrons or as a result of internal conversion. By using a scintillation detector and a pulse-height analyser, the radioisotope can be identified by gamma ray spectrum.

Then irradiate a standard sample containing a known amount of the particular element in exactly the same way as the unknown sample. Again record the gamma-ray spectrum.

By comparing the heights of the peaks called the photopeaks in the gamma- ray spectrum with those of a standard sample, the quantity of the element can be determined.

Sensitivity of the Activation Analysis : The sensitivity of the method is dependent on the atomic cross section, the flux of particles and the half-life of particular nuclide. High cross section and high flux increase the sensitivity. The half-life must be favourable for measurement.

The presence of elements with very high cross-section might interfere when the analysis for some other elements is carried out.

Limits of Estimation : With a neutron flux of $10^{11}cm^{-2}sec^{-2}$, sensitivities of most elements are in the range of 10^{-16} to 10^{-11} gm Thus, activation analysis is used for impurities present in parts per million or even parts per billion.

With a high flux of charged particles, the minimum detectable amount is in the range of 10^{-6} to 10^{-9}gm ; in some instances the minimum is even as low as 10^{-12} to 10^{-16} gm . A well known example is the determination of carbon in steels by means of the reaction C^{12} (d, n) N^{13}.

Advantages of Activation Analysis

(1) It is a non-destructive technique which does not require any removal from the sample to be analysed.

(2) It is highly sensitive and permits the quantitative determination of elements present in such small traces that conventional methods of analysis would fail.

(3) This method provides a simple alternative to much more tedious procedures.

Limitations of Activation Analysis

Like any other method of analysis, it has its limitations. Some of the limitations are discussed below:

(i) ***Elements for which neutron activation is not suitable*** : Almost all elements become radioactive to some extent after irradiation but the half-life of some of the radioactive nuclides formed is too short to allow for removal of the sample from the pile followed by the necessary chemical separation and counting. Elements in this category include He, Li, B, W, O and Ne, with half lives of seconds or less and normal activation techniques are not, therefore, readily applicable to these elements.

Similarly, this technique cannot be used for elements like Al, Mg. Ti, V and Nb because the half-lives of isotopes produced are few minutes.

(ii) ***Physical limitations*** : There are certain physical limitations that one must keep in mind in activation analysis. Heat is being produced and neutrons are also produced during bombardment in a nuclear reaction. Therefore, samples must be of such a character that they can withstand the particular temperature to which they are to be exposed and must be enclosed in such a manner that there is no danger of contaminating the reactor. Liquid must be sealed in silica containers.

(iii) ***Chemical limitations*** : Samples put into a nuclear reactor will be bombarded both by gamma-rays and neutrons. Structural damage and decomposition by process must be considered. Silica, for example, after a few hours irradiation becomes almost black so that one must take into account this effect while performing the activation analysis.

(iv) ***Nuclear limitations*** : The samples must not have a high neutron absorption during irradiation. When cadmium was estimated by activation analysis, this cadmium stopped the chain reaction in the same manner as the control rods of cadmium do in atomic pile.

Another limitation is that the radioactive nuclide which is being used as a measure of the mass of a given element in the activation analysis may be formed from different elements present in the sample to be estimated. For example, the determination of arsenic by reactor irradiation depends on the formation of As^{76} from the monoisotopic As^{75} by the nuclear reaction.

$$_{33}As^{75} + {_0}n^1 \longrightarrow {_{33}}As^{76}$$

However, it is possible as As^{76} can be formed by several other reactions starting with elements present in the sample to be analysed other than arsenic :

$$_{34}Se^{76} + {_0}n^1 \longrightarrow {_{33}}As^{76} + {_1}H^1$$

$$_{35}Br^{77} + {_0}n^1 \longrightarrow {_{33}}As^{76} + 2{_1}H^1$$

Applications of Activation analysis : Several hundred applications have been reported. Some important applications are given below :

(i) Construction materials and moderators for nuclear reactors must he free from elements that capture neutrons to a significant extent only. Activation analysis is a sensitive method for detecting such impurities.

(ii) In the construction of transistors, detectors, etc., the basic materials are usually silicon, germanium, etc., which should be very pure. Traces of undesirable impurities have been estimated by activation analysis.

(iii) Activation analysis has been utilized for estimating traces of undesirable impurities. An example of this is the determination of small amount of oxygen in steel. The oxygen–16 present in the steel is bombarded with fast neutrons, producing nitrogen-16 by the reaction O^{16} (n, p) N^{16}. The oxygen–16 present in the steel is then determined by measuring the gamma rays from nitrogen–16.

(iv) It is interesting to note that trace elements like Zn, Cu, Mn and Co (few parts per million) play an important role in animal and plant life. Their amounts could not be estimated by the usual methods of estimation. However, these have been estimated by activation analysis.

(v) Activation analysis has been used for the analysis of trace elements present in rocks, soils and sands. This study throws light on the origin of the respective materials.

(vi) Activation analysis has been used in dating geological specimens.

(vii) Activation analysis has been utilised for the analysis of trace elements (1 part in 100 million) present in meteorites. This study is useful in determining the ages of these objects that fall on earth from space.

(viii) Recently, Americans have utilised this method for the analysis of compositions of the surface of the moon. This study has provided much useful information about the elements present on the lunar surface.

(ix) Activation analysis has proved very useful in the nondestructive analysis of rare archeological artifacts, where conventional methods have required removal of samples.

(x) Activation analysis has been utilised to identify the geographic origin of specimen of pottery.

(xi) Activation analysis has shown that human hair contains several trace elements in amounts that are characteristic of the individual from whom the hair was taken. It has been suggested, in fact, that the composition of the hair is as specific of a person as his fingerprints. In this connection, the following is of interest. It was reported in 1962 that activation analysis indicated the presence of unusual proportions of arsenic in hair taken from the head of Napoleon after his death on St. Helena. This led to the suspicion that he had been slowly poisoned. Later, however, it was found that hair, removed some years before Napoleon's death contained a similar quantity of arsenic. The poisoning theory is thus open to question. Actually, arsenic is a common trace constituent of hair, although the proportion varies from one individual to another.

(xii) A method, which is much more sensitive than the traditional "paraffin test, "for determining if a person has fired a gun, is based on neutron activation. In the conventional test, molten paraffin is applied to the side of the hand in the region of the thumb and first (trigger) finger; after solidification, the paraffin is removed and tested by chemical methods for nitrates or nitro-compounds normally present in the explosive charge.

3

Natural Radioactivity

INTRODUCTION

Henri Becquerel was studying the phenomenon of phosphorescence, the property of some of the substances to absorb light and re-emit it over a fairly long period of time. One of the substances which he was investigating for this property was a salt of uranium, *viz.* potassium uranyl sulphate. He had kept this salt wrapped in black paper and some photographic film lying side by side in a drawer. He found that the photographic film was blackened. He has kept the film protected from any source of light.

The source of light could be the uranium salt itself. Could it be the phos-phorescent light from the uranium salt that had blackend the photographic plate? This could not be as the salt was not previously exposed to light and was kept wrapped in a black paper. Still there was no source other than the uranium salt. The experimental finding was confirmed more than once. An explanation followed the finding. It was explained that the uranium salt was spontaneously giving off penetrating type of radiation. It was soon found that all uranium salts had this property. Immediatly afterwards the compounds of a number of elements were found to have this property.

According to Becquerel : *"This phenomenon of spontaneous emission of active radiations by certain substances like uranium is called radioactivity while the substances which exhibit this behaviour are said to be radioactive."*

In 1898, Marie and Pierre Curie found that the mineral pitchblende is more radioactive than uranium itself. They suggested that the ore might contain some other element more radioactive than uranium. They started with a ton of pitchblende and working day and night for 3 years, they were able to separate a new element called *polonium* which was more radioactive than uranium.

They continued this work of extraction from pitchblende and were successful in isolating 0.1 gm of another radioactive substance called *radium* which was about million times more radioactive than uranium.

The names radium and radioactivity were coined at the same time. The element then known to give off the most intense of these radiations was called radium and the property called after the element as radioactivity. For their work, Becquerel and Curie shared the Nobel *prize* in physics.

NATURE OF RADIOACTIVE RADIATIONS

It was shown by Rutherford that radioactive radiations consisted of three types. He took a piece of radioactive substance in a cavity bored in a piece of lead metal. Rays from it were passed through a slit and then between two metallic plates connected to opposite poles of a battery. On passing through the electric field, the rays were found to divide themselves into three distinct groups :

(i) Those which deflected towards the negative plate and hence were positively charged. These are called alpha (α) rays.

(ii) Those which deflected towards the positive plate and were negatively charged. These are called beta (β) rays.

(iii) Those which passed undeflected through the electric field and hence were neutral. These are called gamma (γ) rays.

Same results could be obtained by keeping these radioactive rays in a magnetic field.

Alpha Rays

Alpha rays are not rays but consist of positively charged particles moving at high velocity. Therefore, they are called α-particles rather than α-rays. The various characteristics of α-particles are :

(i) *High speed helium nuclei* : α–particles are high speed helium nuclei, shot out from radioactive elements, their velocity being nearly one tenth of the velocity of light; actual velocity depends upon the nature of the radioactive element from which they are obtained.

The charge on an α-particle is $2 \times 4.802 \times 10^{-10}$ e.s.u. This is the same as the charge on the helium nucleus. The ratio of e/m of α-particles was found to be 4.806×104 coulombs per gm.

(ii) *Penetrating power :* The penetrating power of α-rays however, is much less, about 1/100 of that β-rays and 1/1000 of that of gamma rays. A thickness of 0.0005 cm of aluminium foil reduces the ionising power to half and a thick sheet of paper can stop them completely.

(iii) *Ionisation power :* α-Particles have great ionising power, 100 times that due to β-rays and 10,000 times that due to γ-rays.

They produce ionisation in the gas through which they pass, because of their collisions with the gas molecules.

(iv) *Fluorescence :* They produce fluorescence when they fall on certain substances like diamond, zinc sulphide, etc. Diamond fluoresces with blue light whereas zinc sulphide screen gives tiny specks called scintillations. The phenomenon of scintillations was used by Crookes for detecting and counting the number of alpha particles.

(v) *Range of the α-particles :* The absorption of α-particles was studied by Rutherford and Curie in 1903. They showed that α-particles from radium have a range of 3.5 cm. in air at N.T.P.. and after traversing this distance through air, the α-particles lose their power of ionisation as well as the power of exciting fluorescence.

In 1910 **Geiger** showed that the range *R* of particles depends upon the velocity (v) with which they emerge from the source.

$$R \propto v^3$$

or $$R = av^3 \text{ whereas } a \text{ is constant.}$$

(vi) ***Action on photographic plate** : Alpha particles affect a photographic plate and cause luminosity when they strike a zinc sulphide plate. This is due to their high kinetic energy.*

Beta Rays

(i) ***Nature*** : β *rays are negatively charged particles moving with high velocity between* 0.36 *to* 0.98 *times the velocity of light.*

(ii) ***e/m values*** : *The value of e/m, i.e., charge to mass ratio was found to be identical with that of an electron.*

(iii) **Velocity :** *β–particles have tremendous velocities ranging from 33% to 99% of velocity of light. Due to this variation in velocity, beta rays are not homogeneous.*

(iv) **Ionising power :** *The ionising power of β-rays is nearly* 100 *times that of γ-rays but only* 1/100 *times as much as that of α–rays.*

(v) **Penetrating power :** *The β-rays are about* 100 *times more penetrating than α-rays and* 1/100 *times as much as γ–rays.*

(vi) **Effect on a zinc sulphide plate** : *Due to their very low kinetic energy, beta particles a very little effect on a zinc sulphide plate.*

Gamma Rays

(i) **Nature** : *Gamma rays are very short electromagnetic waves shorter than even the hardest X-rays. The wavelength is of the order of* 10^{-10} *cm.*

(ii) **Penetrating power** : *The gamma rays are* 100 *times more penetrating than β–rays and* 10,000 *times more than α-rays.*

(iii) **Ionising power** : *The ionising power of γ–rays is* 1/1000 *times that of β–rays and* 1/100 *times that of α-rays.*

(iv) **Effect of magnetic or electric fields** : *The gamma rays do not show any deviation in a magnetic or electrostatic field, showing that they do not carry any charge but are waves of short wavelength.*

(v) **Effect on zinc sulphide and photographic plate** : *Gamma rays produce very little effect on* zinc sulphide and photographic plates.

(vi) *γ-rays are diffracted by crystals like X-rays, indicating that γ-rays are waves.*

(vii) **Velocity :** *They travel with the velocity of light. Gamma radiation is believed to arise from transitions between energy levels in the nuclei of atoms. These radiations are somewhat analogous to those making up line spectra, which arise from the transitions between the energy levels of the extranuclear structure. Furthermore gamma rays emitted from a given isotope have either the same energy or a discrete set of energies, indicating that nuclear energy levels are probably quantised as are atomic* *energy levels. Therefore, gamma rays can be considered as a nuclear spectrum which may provide information regarding nuclear energy levels in the same manner as optical and X-rays spectra give knowledge concerning electronic energy levels.*

GEIGER-NUTTALL'S LAW

Geiger and Nuttall found in their experiment that, in general, those materials which decay slowly emit α-particles of short range while those which disintegrate rapidly emit more energetic particles. A relationship between the decay constant X, and the range, *R*, was discovered empirically by Geiger and Nuttall in 1921. These are connected with the equation.

$$\log \lambda = A + B \log R \qquad ...(1)$$

where *A* and *B* are constants. This means that for the elements of a particular series *(e.g.,* uranium series) a plot of log X against log *R* will give a straight line, where *R* is the range in standard air.

Relationship (1) is known as Geiger-Nuttall law. This is only approximate. The constant *B* is approximately the same for the three radioactive series, while the constant *A* takes on a different value for each of the series.

STATISTICAL ASPECT OF RADIOACTIVITY

E. **Schweilder** in 1905 made important conclusions about the nature of the radioactivity and redefined it in terms of disintegration probability. His main postulate was that the probability *(p)* for a particular atom of radioactive element to disintegrate in time interval Δt does not depend upon its past history and present circumstances. This probability is proportional to Δt for an extremely short interval. Thus,

$$p \propto \Delta t$$

$$p = \lambda . \Delta t.$$

where λ denotes the proportionality constant. Thus the probability of the atom not disintegrating during this short interval would be given by.

$$1 - p = 1 - \lambda . \Delta t.$$

From the law of compounding such probabilities, the probability for a given atom to survive first interval and also the second is given by $(1 - \lambda . \Delta t)^2$. Thus, for *n* such intervals, the probability would be, $(1 - \lambda, \Delta t)^n$.

The total time, $t = n\,\Delta t$ ...(1)

and the probability= $\left(1-\lambda.\frac{t}{n}\right)^n$

We know that $\lim_{x\to\infty}\left(1-\frac{x}{n}\right)^n = e^{-x}$

Thus, equation (1) reduces to,

$$\text{Probability} = \lim\left(1-\lambda.\frac{t}{n}\right)^n = e^{-\lambda t}$$

If large number (N_0) of radioactive atoms are present initially, the fraction remaining unchanged after time interval t would be.

$$\frac{N}{N_0} = e^{-\lambda t} \qquad ...(2)$$

where N denotes the number unchanged atoms; λ is a constant which is characteristic of radioactive atom and is called *radioactive constant.*

The disintegration of all the radioactive elements is governed by equation (2). The varying values of λ are responsible for the varying radioactive of different elements. But the value of t ranges from milliseconds to million years. Therefore, it is better to know the *average life time* of radioactive element so as to compare their decay.

INDUSTRIAL APPLICATIONS

Radioactive isotopes find several applications in industries where the radionuclide is used as a tracer for identifying a particular element in question, or the radioactivity is made use of to find the location of a material with which the tracer is associated. Or, in some cases, the attenuation or scattering of radiations emitted by the radioisotopes is made use of in the thickness measurement and control in industries. We shall describe some important applications of radioactive tracers below:

A mixture of radioactive thorium and zinc sulphide exhibits a more or less permanent luminescent. This fact has been made use of to use the mixture for coating the pointers and figures of clocks and watches for rendering visible signs.

In the production of metals, these materials are frequently subjected to various different treatments, such as annealing, quenching, cold-rolling, etc. By incorporating radio-isotopes into the metal, it is possible to know what is happening to the constituent elements in the metal subjected to a particular treatment. The distribution of radioactive species in the metals subjected to different treatments can be determined by auto-radiographic method. The use of a radioactive tracer has been made to determine the penetration or diffusion of the radioactive into the metal.

The thickness of coatings or layers, levels of liquids in tanks and autoclaves, or of moving sheets or layers of textiles, paper, rubber sheets, moving on a conveyor belt and controlling their thickness can be measured by using radioactive tracers. The 0.54 MeV-p or Sr^{90} has been used to measure thickness over the range of 60 to 500 mg/cm^2 of matter. In the fabrication of tin plate (*i.e.*, coating of tin on steel), the thickness of the coating can be determined by measuring the backscattering of beta particles.

The extent of backscattering, as observed by a detector on the same side of the sheet as the source, is related to the thickness of the coating. A gamma emitting isotope is used to estimate the liquid levels in a closed tank by following the moving marker technique. The use of radioactive tracer has been made in the transportation of oil in cross country pipe lines, the location of surface of separation between the oil stocks and the extent of corrosion of metallic pipes by liquid circulation.

The tracer technique finds application in some automobile industries for estimating the wearout of piston rings, and their prevention by means of suitable lubricants. Studies involving the mechanism of friction and the effectiveness of various lubricants have been possible by the use of tracer techniques. C^{14} has been used in studying mechanism involved in polymerization, alkylation, catalytic synthesis, catalytic cracking and many other reactions of industrial importance.

Optimum mixing times for materials, such as paint, ink, plastic products, etc., can be detected by using tagged tracer materials. The use of tracers gives important information concerning metallic corrosion as well as inhibition of corrosion.

The metal castings and machine parts for microcracks and defects have been examined by gamma radiography, using y-radiation from certain radioisotopes. Because of greater advantage over the X-ray radiography, y-radiography is becoming a technique of increasing importance in several industries.

The wear of surfaces, *e.g.*, of piston rings and of gears in engines, and its prevention by means of suitable lubricants have been studied in the following manner. The part of interest, such as a steel piston ring, is exposed to neutrons in a nuclear reactor, so that it becomes radioactive. The piston ring is then fitted into the cylinder of an internal combustion engine which is operated in the normal way with a particular lubricating oil.

By determining the radioactivity removed by the oil, the exent of wear of the piston can be determined. Furthermore, if a photographic film is placed against the walls of the cylinder, after cleaning off the oil, the radioactive material transferred from the piston ring Can be detected by its radioautograph. Improvements in lubricating oils, and consequently in piston ring and cylinder wear, result from such obseervaions.

Metals in various f&ms, particularly as alloys of several elements, are frequently subjected to different treatments, such as agehardening, annealing, quenching, and cold-rolling. In order to understand the effects of such treatments, it is desirable to know what is happening to the constituent elements. Radioactive isotopes of these elements provide a simple and effective tool for followig their behavior, for the radioactive atoms will behave in exactly the same manner as the stable atoms of a given element. The location of the radioactive species can be determined at any time during the treatment of the alloy by making a radioautograph.

The phenomenon of self-diffusion in a metal, *i.e.*, the movement of the atoms of a metal within the crystal lattice, is of interest in connection with the properties of metals under stress at high temperatures. Since it is not possible to follow the diffusion among identical atoms, there is no way, wihtout the use of a radioactive tracer, to obtain experimental data on this subject.

A block of metal, such as copper, consisting of the stable isotopes, has a layer of the same metal containing a radioisotope deposited on it, *e.g.*, by electroplating. After subjecting the block to heat treatment, stress, etc., successive thin layers are shaved off the surface and their activity measured. In this way, the penetration or diffusion of the radioactive into the stable metal, which is the same as ordinary self diffusion, can be determined.

Metallurgists have long been interested in the sulfur present in coal; some of this remains in the coke made from the coal, and a part ultimately finds its way into steel, where it is not desired. In one particular instance it was required to know whether it was organic sulfur in the coal or inorganic sulfur, usually present as the mineral pyrite, which remained in the coke. The problem was readily solved by adding to the coal a small quantity of pyrite containing some radioactive sulfur, which could be traced into the coke. Since the stable and radioisotopes of sulfur in the pyrite behave alike chemically, the results provided the desired information. It was found that the proportion of organic to inorganic sulfur in the coke was the same as that in the coal from which it was made.

Another practical example of the use of radioisotopes in process metallurgy originated in the attempt to determine whether a fine powder of iron ore, obtained by concentration from a low-grade source, could be charged into a blast furnace or whether it would be blown out by the air blast. A quantity of the fine ore was exposed to neutrons in a reactor, so that it became radioactive. This was mixed with a large amount of the original fine ore and then with a different ore of a coarser type. After treatment in the blast furnace, the pig iron, slag, and dust were examined for radioactivity. The results showed that 60 percent of the fine ore remained in th furnace as pig iron. Although this proportion was higher than expected, the loss was too large to be tolerated. The results showed that further studies are needed to determine how to increase the retention of the fine material.

In the preceding example, the purpose of the tracer was to follow the movement of a bulk quantity of material. Although part of the ore was itself made radioactive by exposure to neutrons, the required information could probably have been obtained by using a tracer that was not directly involved in the process taking place. There are many instances of applications of this type in industry. For example, in the decomposition ("cracking") of petroleum to produce gasoline, the powdered catalyst is used in the form of a fluidized bed. For maximum efficiency, the rate of flow of must be controlled. This is frequently done by mixing a small quantity of a gamma-emitting radioactive nuclide, *e.g.*, scandium-46 or zirconium-95, with the catalyst particles. Small samples are withdrawn at various locations and their radioactivity determined with suitable counters, can be related to the flow rate of the catalyst. This procedure has also proved valuable in reducing loss of the catalyst powder through the stacks.

Sponge iron is often produced in a continuous process involving the reduction of iron ore pellets in a furnace. In order to follow the movement of these pellets, spheres of the same size were made of a refractory material and a small picee of wire containing radioactive iridium-192 was embedded in each sphere. Three sets of spheres were prepared with different levels of radioactivity in order to permit their identification. One set of these labeled (tracer) spheres was mixed with the ore pellets near the wall of the furnace, another near the center, and the third set in between. The times at which the different tracer spheres left the furnace provided information concerning the flow of the ore in the reduction process.

There are many instance in which a radioactive substance is used to follow the motion of liquids and gases, as well as of solids. Uniformity of mixing during the blending of gasolines, lubricating oils, greases.etc., can be readily observed if one of the constituents is labeled with a radioactive tracer. When the mixing is complete, samples taken from different locations will have the same activity. Mixing of liquids in surge tanks and stills can be followed in a similar manner. Radioactive tracers also provide a convenient means for detecting leaks, especially in buried pipes carrying water or petroleum. A small quantity of a radioactive substance is dissolved in the liquid near the point of the suspected leakage. The actual location of the leak can then be found by means of a sensitive gamma-ray counter, although the escaping liquid is not visible.

Several methods of industrial gauging, *e.g.*, of the thickness of a material or the level of a liquid in a tank, utilize the absorption or scattering of radiation from a radioactive source. For checking the thickness of sheets of paper, cellophane, plastic, and rubber, and even of metal (steel) plates or pipe, a source of radiation is placed on one side and a detector on the other side. If the material to be gauged is paper or other fairly weak absorber, the radiation can be beta particles, but for steel it is necessary to use gamma rays.

The proportion of the radiation absorbed, and hence the amount reaching the detector, depends on the thickness of the intervening material through which the radiation passes. A device of this kind gives a continuous record of the thickness while the machine is operating. It can be adapted to automatic control by providing an instantaneous signal that can be utilized to change the setting of the rollers in the production of sheets.

When one material is coated on another, *e.g.*, tin on steel in the fabrication of tin plate, the thickness of the coating can be determined by measuring the backscattering (or reflection) of beta particles. The extent of backscattering. as observed by a detector on the same side of the sheet as the source, is related to the thickness of the coating. A similar technique has been used for checking the thickness of the asphaltic layer applied on road paving.

Radiation is also employed as a means for indicating the level of a liquid in a closed tank. One way in which this is done is to place a gamma-ray source outside the tank on one side and a detector at the same level on the other side. If the source- detector level is below that of the liquid in the tank, much of the radiation will be absorbed in its passage through the tank and the detector will indicate a low reading. On the other hand, if the source-detector level is above that of the liquid, the gamma rays will pass through air (or vapour) and a much higher reading will be observed. Consequently, if the source and detector are moved vertically, the level of the liquid in the tank is indicated by an abrupt change in the detector reading.

Information concerning the location of the surface of separation, the velocity of flow, and the degree of intermixing of oil stocks flowing in a pipe-line can be obtained by the use of a radioactive indicator injected between two different oil stocks. The 12.8-day barium-140 was originally employed for this purpose, but on account of its short life it lost a large proportion of its activity during the preparation and transportation of the material. Consequently, antimon-124, with a half-life of 60 days, is now commonly used ; this nuclide emits , in addition to beta particles , gamma rays of long range which can be detected by means of a counter placed outside the pipe-line. Thus, the position and sharpness of the interface between the oil stocks can be determined. The decay product of antimony-124 is a stable isotope of tellurium, and hence the radioactivity disappears in a short time, so that it does not constitute a hazard.

AGRICULTURAL APPLICATIONS

Radioactive isotopes have been widely used in agriculture to improve the plant growth. A very important question now arises. At what time in the plant's life it needs radioisotopes? In what form and to what extent the radioisotopes are necessary ? We shall answer these questions by considering the use of some important isotopes in various aspects of agriculture.

Radioactive P^{32} is added with phosphate fertilizer to the soil to improve plant growth. The uptake of P^{32} from the soil by plants is determined by measuring the radioactivity at different parts of the plants.

The total amount of phosphorous taken up by the whole plant is determined by chemical analysis and that of the added fertilizer by the activity measurement. The difference is the natural phosphorus present in the soil. Thus, the use of P^{32} as a tracer provides a means of ascertaining the kind of phosphate which is efficient for a given soil and crop and also the stage of plant growth at which the fertilizer to be added. By using various P^{32}-labeled phosphates it has been shown that ammonium phosphate is a much more effficient fertilizer than superphosphate or calcium phosphates. But, nevertheless, the kind of phosphate fertilizer depends on the type of soil, its phosphorus content and the nature of the crop. In general, the higher the amount of phosphorus in the soil, the larger is the proportion which the plant takes up from the source. The action of phosphate fertilizer takes place particularly during the early stages of growth of plants, when about 65% is taken up but in the later stage , most of the phosphorus comes from the soil.

It is also possible to investigate the relationship between root growth and the phosphorus uptake by placing fertilizer labeled with P at different depths and distances from growing plants. Radioisotopes have provided a means of studying the feeding of plants through their leaves. The radioactive carbon (C^{14}) has been helpful in understanding the mechanism of photosynthesis in plants. This process ivolves the formation of sugar and starches in the presence of sunlight and the green leaves (chlorophyll) by the interaction of carbon dioxide and water. By using $C^{14}O_2$ with $C^{12}O_2$, it has been shown that oxygen which is formed along with sugar, comes from water and not from carbon dioxide.

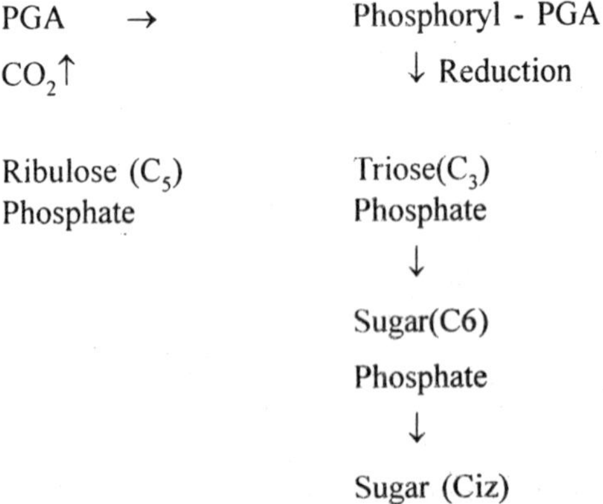

Fig. 1. Outline of the path of carbon in photosynthesis.

Radioactive iron tracer has been used to investigate the disease, chlorosis developed in the plant because of the shortage of chlorophyll. Isotopic tracers have also been used in studying animal nutrition and metabolism. Many fungicides contain sulphur. The use of S^{35} tracer indicates the advantages and disadvantages caused by these fungicides. Radioactive isotopes are being used in various aspects of agriculture to provide information that could not be secured in any other way.

For example, phosphorus is often applied to the soil to improve plant growth, but there are several types of phosphate fertilizers available and it is not always obvious which is utilized most effectively by a given type of soil. The total amount of phosphorus taken up by the plant can be determined by ordinary chemical analysis; but before the advent of tracers there was no way of distinguishing between the phosphorus derived from the soil and that obtained from the added fertilizer. The use of radiophosphorus as a tracer permits a distinction to be made and provides a means of ascertaining the kind of phosphate which is best for a given soil and crop.

The procedure adopted is to start with phosphoric acid containing a known proportion of the radioactive phosphorus-32, and to convert this into a phosphate, such as calcium superphosphate, tricalcium phosphate, or hydroxyapatite, suitable for use as a fertilizer. A definite quantity of the labeled phosphorus is then applied to the soil in which plants are grown. At certain intervals a number of these plants are harvested, and the total phosphorus, derived from both soil and fertilizer, is determined by chemical analysis of their ash. Assuming, as is very probable, that the plant does not distinguish between ordinary and radioactive phosphorus atoms in the fertilizer, the radioactivity of the plant ash, combined with the measured specific activity of the fertilizer, immediately gives the amount of phosphorus which the plant takes up from the latter. In this way the individual quantities derived from the soil and from the fertilizer can be estimated.

Because experiments of this type, which depend on the growth of plants, require a considerable time, the data are accumulated relatively slowly. Nevertheless, a number of interesting points have become apparent, and a few of them will be mentioned here. The proportion of phosphorus absorbed from a given soil and fertilizer depends on the nature of the plant and on the period of growth. Tobacco and corn, for example, both derive about 65 percent of their phosphorus from the fertilizer in the early stages of growth, but near maturity the proportion has dropped to 45 percent for the former, and to only 15 percent for the latter. Evidently, in this particular, rather poor, soil, corn requires most of its phosphorus fertilizer at the beginning, and less toward the end of its growth. With potatoes on a similar soil, from 50 to 60 percent of the phosphorus uptake came from the fertilizer during the whole period of growth.

The ratio of phosphorus derived from the soil to that obtained from the fertilizer varies with the type of soil, its phosphorus content, the amount of fertilizer added, and the nature of the crop being grown. In general, the higher the amount of phosphorus in the soil, the larger is the proportion which the plant obtains from this source. In greenhouse experiments made in Canada, in which wheat plants were grown in soil to some of which ammonium phosphate was added as fertilizer, it was found that the plants actually took up more phosphorus from the soil in the presence of the fertilizer than they did in its absence.

A study of great interest was made of the availability of phosphorus in green manure. Wheat fertilizer with labeled phosphate w+s grown as a green manure crop; it was then analyzed for its total phosphorus content and mixed with soil so as to give the equivalent of 65 pounds of phosphoric acid per acre. To another quantity of the soil was added an equal amount of phosphoric acid in the form of superphosphate, also labeled with phosphorus-32. Rye grass was planted in both cases, and from the radioactivity of the crops it appeared that the green manure phosphate is nearly as effective in supplying phosphorus to the plant as is superphosphate.

The relationship between root growth and the uptake of phosphorus from the soil has been investigated by placing fertilizer labeled with radiophosphorus at several depths and distances from growing plants. After various time intervals, the plants were harvested and analysed for radiophosphorus. The results provided information on the manner in which the phosphorus is taken up by the roots. With corn and certain other plants, for example, it was found that, as the plant develops, the root feeding system is below the usual depth at which phosphate fertilizer is placed in the soil. Hence, when the plant was beginning to mature it was getting no advantage from the added phosphorus. Somewhat deeper location of the phosphate gives better yields of these plants.

Radioisotopes have provided a unique method for studying the feeding of plants through their leaves. It has been shown, with nitrogen, phosphorus, and potassium tracers, that nutrients are frequently absorbed much more readily from the leaves than through the root system. Absorption can also take place through the twigs, flowers, and even the fruit. Although, chlorophyll contains no iron, its formation in the green plant requires the presence of iron; if there is a deficiency of this element in the soil, plants develop the disease called chlorosis. The leaves are then yellow, instead of green, because of the shortage of chlorophyll, and photo-synthesis is greatly reduced. Plants sometimes exhibit chlorosis in certain soils which appear to contain adequate quantities of iron. The cause of this behavior was discovered by using a radioactive non-tracer, it was found that certain elements can interfere with the absorption of iron by the plant.

Information of this kind has been applied in the successful growth of cotton plants, which was previously not possible, in the Tulare Lake region of CaUfornina. By radioactive tracer techniques it was esstablished that the plants needed zinc for adequate growth, but the uptake of this element was inhibited by phosphorus, which is esential to all plants, that is present in the soil. If extra zinc was added, however, the combination of zinc and phosphorus interfered with the absorption of iron and chlorosis developed in the plants. By includig zinc and iron together with phosphorus in the fertilizer, excellent crops of cotton plants were produced.

Isotopic tracers have also been used in several studies of animal nutrition and metabolism. For example, in order to determine the quantity of calcium taken up by a cow from a particular feed, the latter is labeled with radiocalcium. By observing the radioactivity of the milk and the excretions, the fate of the calcium in the food can be traced. Similar studies can be made with radiophosphorus. These have shown that most of the phosphorus in a cow's milk comes from the element already in the bone and a relatively small proportion is derived directly from the feed.

It was mentioned that measurements of the total amount of natural radioactive potassium-40 in the human body, with a whole-body counter, can serve to indicate the relative proportions of fat and lean tissue. Similar estimates can be made for animals. By means of this technique it is possible to determine, with living animals, the best diet for encouraging the developement of lean meat, rather than of unwanted fat, in a steer.

APPLICATIONS OF NUCLEAR RADIATIONS

So far we have discussed the use of radioisotopes as tracers in various aspects of science. We shall now discuss the direct applications of radiations in the process of food preservation, sterilization, chemical synthesis, scintillation counters, etc.

Food Preservation by Radiation

It is a fact that all foodstuffs, vegetables, milk, eggs, and meat have a limited life after which they are spoiled and become inedible. This is caused by the action of various micro-organisms, such as bacteria and molds, when the number of micro-organisms exceeds $10^8 g^{-1}$ of matter. An exposure of the foodstuffs to an appropriate dose of eradiation from a Co^{60} (or possibly Cs^{137}) source at room temperature can destroy the harmful micro-organisms so that the shelf-life (or storage time) of the foodstuffs is increased without changing the taste, odour, texture and appearance. There is always an optimum dose for each material, otherwise with an excess of dose, there is destruction of vitamin contents in the food and spoilage of food becomes faster. Thus a dose of 50-200 Gy given to potatoes prevents sprouting for a few months longer than unirradiated ones. Water can be adequately sterilized by a γ-dose of the order of 2 kGy.

Sterilization by Radiation

Surgical instruments, such as sutures, gloves and ampules are sterilized by exposure to gamma radiation from Co^{60} replacing the conventional way of steaming in an autoclave or by chemicals. Sterilization by radiation helps the materials to retain the strength and pliability.

Chemical Effects of Radiations

A large number of chemical reactions initiated by radiations are known to occur at room temperature without a catalyst giving rise to purer products. Some of the reactions having actual applications in industry are described below :

Ethyl bromide is an important intermediate in the large-scale synthesis of organic compounds. Ethylene and hydrogen bromide gases are passed into liquid ethyl bromide (prepared from ethyl alcohol and HBr gas) and this liquid is allowed to pass through the irradiation zone at room temperature and on exposure to a high intensity Co60 source gives rise to ethyl bromide of 99.5% purity. The yield, GC_2H_5Br) comes to of the order of 106.

Gammexane ($C_2H_5Cl_6$) is prepared by chlorination of benzene on exposure to γ-radiation with a yield, $G(C_2H_5Cl_6)$ of the order of 10^5. The production and modification of plastics are also possible using nuclear radiations.

The use of radiation-induced polymerization in the production of irradiated wood-plastic combinations has been made to increase the strength of wood and to make it the water absorption resistant. Radiations have been used to control the insect pests in the plant crops and livestocks, and in various aspects of science.

Autoradiography

The interaction of radiation with photographic plate results in fogging or blackening of the film. This fact has been made use of for locating the position, transport, or dispersion of a material within a complicated matrix.

However, with a nuclear emulsion of a given composition, the blackening is greatest where the radiation is most intense, so that the points of emission can be localized, and the relative intensities of the radiation obtained by microphotographic methods.

CAUSES OF RADIOACTIVITY

Since radioactivity is a nuclear phenomenon, it must be connected with the instability of the nucleus. Since all elements are not radioactive, the ratio of neutron to proton of the unstable radioactive nucleus, is the factor responsible for radioactivity. It has been reported that the stability or the instability of the nucleus is connected with the pairing of spins of neutrons among neutrons, and pairing of spins of protons among protons. When all the non-radioactive isotopes of the elements are considered, it is found that nuclei with even number of protons and with even number of neutrons are most abundant. It will be noted that even numbers lead to spin pairing, and odd numbers lead to unpaired spins. *Nuclei with either of the proton number or neutron number as odd are slightly less stable than even numbered ones. But the least stable isotopes are those which have odd number of protons and odd number of neutrons.* The nuclear spin pairing is a minimum when odd numbers of both are present.

Furthermore, the nuclear stability is also mainly influenced by the *relative numbers* of protons and neutrons in addition to odd or even numbers of nucleons.

The neutron/proton ratio (n/p) helps us in predicting in which unstable radioactive nuclei will decay. A plot of number of neutrons in the nuclei of the stable isotope against the respective number of protons. A study of the figure reveals that the actual *n/p* plot of the stable isotopes breaks off from the hypothetical *n/p* (1 : 1) plot at around an atomic number 20, and thereafter rises rather steeply. This indicates that as the number of protons increases inside the nucleus, more and more neutrons are needed to minimise the proton-proton repulsion and thereby the nuclear stability increases. Neutrons, therefore, serve as binding material inside the nucleus. The way an unstable nucleus will disintegrate will be decided by its position with respect to the actual n/p plot of stable nuclei. When an isotope is found above this plot it has too high a *n/p* ratio, and when it is located below the plot it has too low *n/p* ratio. In either case the unstable nucleus should decay so as to approach the actual *n/p* plot. We now explain the two cases of decay :

1. Neutron-to proton too high. An isotope with too many neutrons inside the nucleus (that is, with more neutrons than is needed for stability) can attain greater nuclear stability if one of the neutrons decays to a proton. Such a disintegration leads to the ejection of an electron from inside the nucleus.

$$_0n^1 \longrightarrow {}_1H^1 + {}_{-1}e^0$$

Thus, beta ray emission will occur whenever the *n/p* ratio is higher than the value expected for stability.

The *n/p* ratio for carbon-12 is 1.0 but that of carbon-14 is 1.3. It can be predicted that carbon–14 will be radioactive and will emit beta rays.

$$_6c^{14} \longrightarrow {}_7N^{14} + {}_{-1}e^0$$

2. Neutron to-proton ratio too low. A nucleus deficient in neutrons will tend to attain nuclear stability by converting one of its protons to a neutron and this will be achieved either by the emission of a positron or by the capture of an orbital electron. Such a decay occurs with a radioactive isotope whose masses number is less than the average atomic weight of the element.

Positron emission occurs with light isotopes (low *n/p)* of elements of low atomic number. Thus, nitrogen-13 (*n/p* = 0.86) decays by positron emission.

$$_7N^{13} \longrightarrow {}_6C^{13} + {}_{+1}e^0$$

Orbital electron capture occurs with isotopes (too low n/p) of elements of relatively high atomic numbers. In such cases die nucleus captures an electron from the nearest orbital (*K* shell; *n* = 1) and thus changes one of its protons to a neutron. For example :

$$_{79}A\mu^{184} + {}_{-1}e^{0} \longrightarrow {}_{78}Pt^{194}$$

The net effect of the nuclear transformation resulting from orbital electron (*K* electron) capture is the same as that observed with positron emission. In both the cases, a proton in the nucleus is converted into a neutron. The bismuth-209 (atomic number 83) only, since no nucleus heavier than this is known to be stable. Thus, we conclude that among the heaviest nuclei the total proton-proton repulsion is much too large, *i.e.,* so large that the binding effect of neutrons is not enough even to lead to a single, stable non-radio-active isotope. For such nuclei alpha particle emission is a common mode of decay. For example :

$$_{90}Th^{112} \longrightarrow {}_{88}Ra^{228} + {}_{2}He^{4} : {}_{92}U^{238} \longrightarrow {}_{90}Th^{234} + {}_{2}He^{4}$$

The isotopes $_{88}Ra^{228}$ and $_{90}Th^{234}$ are not stable either. These decay till the decay products are isotopes of stable, non-radio-active lead.

RADIOACTIVE DISINTEGRATION SERIES

Whenever a radioactive disintegration occurs with the emission of an alpha or a beta particle, the original atoms, called *parent,* change into something else, called the *daughter.* In 1950 Rutherford and Soddy proposed that the nature of the daughter could be inferred from the nature of the parent and the particle emitted. We can state Rutherford Soddy rules as :

1. The total electric charge (atomic number) or algebraic sum of the charges before the disintegration must be equal to the total electric charge after the disintegration.
2. The sum of the mass numbers of the initial particles must be equal to sum of the mass numbers of the final products.

There are four main series of radioactive elements : the *thorium, uranium, actinium and neptunium* series. The first three natural disintegration series, are sometimes referred to as the 4n (thorium), 4n + 2 (uranium) and 4n + 3 (actinium) series. In all the series, a parent radioactive element of large atomic mass and very long life gives rise to a series of radioactive elements as a result of the successive emission of α of β particles. The transformation may be represented by the expression in the form

$$_{Z}X^{A} \longrightarrow {}_{Z-2}Y^{A-4} + + \alpha$$

$$_{Z}X^{A} \longrightarrow {}_{Z+1}Q^{A} + \beta$$

The daughter products Y and Q are themselves radioactive if they have atomic weights greater than 82, the atomic number of lead.

General Characteristics

Thorium Series (or 4n series) : This series is named after the longest lived nuclide of thorium, Th-232 (half life = 14.000 million years). The series ends with the stable $_{82}Pb^{208}$ isotope. The atomic masses of all members of this series are exactly divisible by 4; hence the name "4n series". The mode of decay is presented in Table 1.

Uranium Series (or 4n + 2 series) : Beginning with longest lived $_{92}U^{238}$ (half-life = 4,500 million years) and ending up with the stable $_{82}Pb^{206}$.

Actinium Series (or $4n + 3$ series) : Derives its name not from the longest lived nuclide (U–235 having a half-life of 713 million years), which occurs in natural uranium to an exceedingly small extent, but form the nuclide, Ac–227, which is the *sole constituent* of natural occurring actinium. The atomic masses of all the members of this series correspond to the expression $(4n + 3)$.

Some of the general characteristics of the natural series are;

(i) All the three radioactive series have been named after a prominent member in each decay series.

(ii) Radioactive decay chain is not always straight-forward. There occurs certain branches into the decay chain, wherein element undergoes a dual decay, *i.e.*, it gives off α as well as β particles.

(iii) An isotope in one series does not decay to a particular isotope belonging to another series.

(iv) The actinium series was at one time thought to originate with this element, but it has since been discovered that the true parent to the series is a much longer-lived element.

(v) There is a fairly close similarity among the three series and their modes of decay. Originally, the members of the decay series were designated by names such as Uranium X_1, Uranium I, Uranium X_2 etc., and thorium, mesothorium I, etc.

The historical names are now becoming obsolete and shall refer to the ratio isotopes in terms of the element to which each belongs.

(vi) The ultimate product in each of the three series is a stable, but different in each case, isotope of lead. Mass numbers of lead isotopes of the uranium, thorium and actinium series are 206, 208 and 207 respectively, although the actual masses of lead isotopes have been determined to be 206.5 208.05 and 207.05 respectively.

No Specific Relation. These three series are not specifically related. An isotope in one series will not decay to particular isotope belonging to another series.

Neptunium Series (or $4n + 1$ series) : With the discovery of nuclear fission and production of neptunium another series was also found which starts with plutonium 241 and ends up with the element bismuth-209 which is stable. The atomic masses of the members of this series are expressible by $4n + 1$. This is the *artificial series* of nuclides. The half-life of the longest-lived nuclide of neptunium series is neptunium itself with half- life of 2.2×10^6 years.

Table 1

Element	*Nuclide*	*Radiation*	*Half-life*
Plutonium	$_{94}Pu^{241}$	β	14 y
Americium	$_{95}Am^{241}$	α	470 y
Neptunium	$_{93}NP^{237}$	α	2.2 ×'106 y
Proto-actinium	$_{91}Pa^{233}$	β	27.4 *d*
Uranium	$_{92}U^{233}$	α	1.62 × 105 *y*
Thorium	$_{90}Th^{229}$	α	7340 *y*
Radium	$_{88}Ra^{225}$	β	14.8 y
Actinium	$_{89}Ac^{225}$	α	10 *d*
Francium	$_{87}Fr^{221}$	α	4.8 *m*
Astatine	$_{85}At^{217}$	α	0.018 s
Bismuth	$_{83}Bi^{213}$	β, α	47 *m*
(Polonium (96%)	$_{84}Po^{213}$	α	4.2×10^{-8} s
(Thallium (4%)	$_{81}Tl^{209}$	β	2.2 m
Lead	$_{82}Pb^{209}$	β	3.3 *h*
Bismuth	$_{83}Bi^{209}$	stable	stable

In addition to this, artificial α-emitters have been obtained which make a part of either of the naturally occurring series, *i.e.*, they run parallel to the part of these series. These constitute the so called *collateral series.* For instance there is a series, starting with U^{233} which is collateral to the actinium series.

This series differs in several respects from the naturally occurring radioactive series.

(i) The end product in the Np-237 series is the stable isotope of bismuth $_{83}Bi^{209}$ whereas the end products of the three natural radioactive series are stable isotopes of lead.

(ii) The only member of the series to be found in nature is the stable end product, bismuth $_{83}Bi^{209}$

(iii) The series contains no gaseous emanations as do the three natural series.

(iv) Branched disintegration appears to be more frequent in occurrence into naturally radioactive series than in the neptunium series.

NUCLEAR REACTIONS

Introduction : In ordinary chemical reactions, the nuclei of the atoms taking part in a chemical reaction, remain unaffected and only the electrons in the extra nuclear part of atoms take part in the chemical process.

However, during disintegration of atoms (naturally or artificially), the nuclei of atoms are affected resulting in the formation of new nuclei : *Such reactions, in which the nuclei of the atoms interact with other nuclei or lighter particles or photons, resulting in the formation of new nuclei and one or more lighter panicles are called nuclear reactions.*

Following facts are taken into account while expressing a nuclear reaction :

(i) ***Reaction written like a chemical equation*** **:** *Nuclear reactions are written like a chemical equation. Reactants are written on the left hand side and products on the right hand side with an arrow in between.*

(ii) ***Mass number and atomic number of elements involved are also written*** **:** *Mass number is written as superscript and atomic number as subscript on the symbol of the element. For example,*

$_7N^{14}$ *stands for an atom of nitrogen with mass number* 14 *and atomic number 7*

(iii) ***Mass number and atomic number are conserved*** : *As in a chemical reaction the total number of atoms of various elements is balanced on the two sides, similarly here the total mass number and atomic number are balanced on the two sides.*

(iv) ***Symbols used for projectiles*** : *Like symbols for the atoms of the elements, projectiles are also represented by following symbols:*

$_0n^1$	for neutron
$_1H^1$ or p	for proton
$_2He^4$ or α	for α particle
$_{-1}e^0$ or e	for electron or β-particle
$_+1e0$	for a positron
$_1H^2$ or $_1D^2$	for a deuteron

(v) ***Energy evolved is written as + ϕ or energy absorbed as -ϕ on R.H.S*** : *Like chemical reactions, nuclear reactions are also accompanied by release or absorption of energy. This is represented by adding ϕ on the R.H.S. Thus the nuclear reaction between Al and a particle α may be represented as*

$$_{13}A^{27} + {}_2He^4 \longrightarrow {}_{15}P^{30} + {}_0n^1 + \phi$$

Similarly, interaction between nitrogen and a-particle is written as

$$_7N^{14} + {}_2He^4 \longrightarrow {}_8O^{17} + {}_1H^1 + \phi$$

(vi) ***Short hand method*** : *Sometimes a short hand notation is also used. In this form above reactions are represented as :*

Al^{27} (α, n) P^{30} and N^{14} (α, p) $_8O^{17}$ respectively.

Thus, in this notation, the nucleus bombarded is written first then within brackets, the projectile and he lighter particles produced and finally the new nucleus produced is written as shown above.

Energetics of Nuclear Reactions : As energy and mass are interconvertible inside the nucleus the total amount of energy in nuclear reactions generally includes the rest energy.

Suppose the energy of a target nucleus be M, and that of a projectile be $M_p + E_p$. Due to the interaction of these two suppose a nucleus of energy $M_b + E_b$ and a particle of energy $M_n + E_n$ are produced. According to the law of conservation of energy, we have

$$M_t + M_p + E_p = M_a + E_a + M_b + E_b \quad \text{... (1)}$$

$$(M_t + M_p) - (M_a + M_b) = E_a + E_b - E_p \quad \text{... (2)}$$

If ΣM_i and ΣM_f be the initial and final masses respectively, then we get

$$\Sigma M_i = M_t + M_p$$

$$\Sigma M_f = M_a + M_b$$

Similarly, $SE_i = E_p$. initial energy

Substituting these in Eq. (2), we get

$$\Sigma M_i - \Sigma M_f = \Sigma E_f - \Sigma E_i = -(\Sigma E_i - \Sigma E_f) \quad \text{...(3)}$$

Thus, $Q = \Delta M = -\Delta E$

where Q is *energy of reaction.*

The equation (2) may be put as follows :

$$M_t + M_p = M_a + M_b + Q \quad \text{... (4)}$$

If $Q > 0$, the reaction will be *exoergic* and for $Q < O$, the reaction will be *endoergic.*

Nuclear Reactions Versus Chemical Reactions : Nuclear reactions differ from chemical reaction in the following respects :

1. Chemical reactions involve some loss, gain or overlap of outer orbital electrons of the reactant atoms. On the contrary nuclear reactions involve emission of alpha particles, beta particles or positrons inside the nucleus. Nuclear reactions lead either to the birth of another element or product isotope of the parent element.
2. The nuclear reactivity of a radio element is independent of its state of chemical combination. The radium atom in elementary radium and the radium ion in $RaCl_2$ are similar in their radioactive behaviour.
3. A chemical reaction is balanced in terms of mass only but any nuclear reaction must be balanced in terms of both mass and energy.
4. Nuclear reactions are accompanied by energy changes which far exceed the energy changes in chemical reactions. In chemical reactions the energy is expressed in terms of kilojoules per mole

while in case of nuclear reactions it is expressed in MeV (million electron volts) per individual nucleus.

5. The chemical reactions are dependent on external conditions such as temperature and pressure, where nuclear reactions are not.
6. In expressing chemical reactions we take into consideration the number of extra nuclear electrons. In nuclear reactions this is not considered since nuclear reactions take place within the nucleus.

Models for nuclear reactions. The data collected from the nuclear reactions can be explained on the basis of three models such as optical model, Bohr's compound nucleus model and direct interaction model. We shall explain these models one by one.

1. Optical model. The potential energy well may be used in this model to display the interaction between the nucleons of the nucleus and incident particle. This model is considered analogous to the emergence of a light beam from a transparent glass ball. In this model, the path of the incident beam gets deflected from the original direction due to nuclear interactions. From this model, the following calculations can be made:

(a) Cross section of the deflection of the incident particle,

(b) Angular distribution of this deflection,

(c) Cross section of the absorption of the incident particle.

The main drawback of this model is that the fate of the particle after absorption is not known.

2. Bohr's Compound Nucleus Theory of Nuclear Reactions. In 1936 Bohr produced his theory of compound nucleus which has been proved helpful in the correlation and interpretation of nuclear reactions. According to Bohr, the nuclear reaction is a process, involving two steps

(i) **First Step.** The projectile combines with the nucleus to form a compound nucleus. In this process the kinetic energy of the projectile is distributed uniformly among all the nucleons (protons, neutrons) in the nucleus and during this state the nucleus is said to be in the excited state.

$$\text{Incident particle} + \text{initial nucleus} \longrightarrow \text{compound nucleus}$$

(ii) **Second Step.** The compound nucleus formed in first step breaks up to give the final products.

Compound nucleus ⟶ product nucleus + out going particle,

It is observed that the Bohr's assumptions are in accordance with many of the factors of nuclear transmutation. When Al^{27} is bombarded with protons, the new nuclide may be Mg ^{24} Si^{27} or Na^{24}. Bohr assumed that the interaction between A^{27} and a proton is assumed to give the compound nucleus $[_{14}Si^{28}]^*$ which may disintegrate in any one of several ways :

Discussion. Bohr' assumption is also in accordance with picture of the nucleus in which it is regarded as a system of particles held together by very strong short range forces. As soon as the particle is incident on the nucleus, it enters the nucleus and its energy is quickly shared among nuclear particles before re-emission can occur. The compound nucleus so formed possesses the following characteristics :

(i) *State of compound nucleus is independent of the way it was formed. For example, the compound nucleus may be formed in any one of the following ways :*

(ii) *The life-period of the compound nucleus is found to be of the order* 10^{-12} *to* 10^{-14} *sec. This time is long compared with natural nuclear time. During this time, the compound nucleus forgets how it was formed and its disintegration is independent of the mode of formation.*

This compound nucleus is in quasi-stable state which means that it has its existence for such a long interval of time but it disintegrates after emitting one or more nucleons. This quasi-stationary state is referred to as ***virtual level*** **or** ***virtual state.***

(iii) *The products formed and the amount of energy released depend upon the energy of the compound nucleus and the manner in which it has been formed.*

Success : With the help of Bohr's theory, the nature of products in a given reaction can be predicted with certain limitations.

3. Direct interaction model. This model predicts as to what happens to the incident particle after absorption and differs from compound nucleus model in that the energy of the incident particle is randomly distributed among the nucleons of the target nuclei. It is presumed in the model that the incident particle interacts with one or some particles in the nuclei and some of them may directly be ejected. Another possibility is that the incident particle may lose some of its energy in such an interaction and leaves the target. It is, therefore, definite that no intermediate excited nucleus is formed and it is expected that the :kinetic energy of the emitted particles would be greater than that of the particles ejected from the excited compound nucleus.

This mode includes such incidents in which a small portion of the complex particle (*e.g.,* deuteron) strikes the target. This is known as *stripping reaction.* The reverse of the stripping reaction, known as *pick-up* process, is also observed. In this process a complex particle (D^2 or He^3) is formed due to the nteraction of incident particle (proton) with nucleon or a group of nucleons.

Types of Nuclear Reactions. Nuclear reaction may be classified under two headings :

1. In terms of the over all energy transformations. Under this, one can distinguish the following types of reactions :

(a) Capture reactions. In these reactions, the bombarding particle is absorbed with or without the emission of gamma radiation. Neutron capture is very common.

$$_6C^{12} + {_1H^1} \longrightarrow {_7N^{13}} + \gamma$$

$$_{35}Br^{74} + {_0n^1} \longrightarrow {_{35}Br^{75}} + \gamma$$

$$_{92}U^{238} + {_0n^1} \longrightarrow {_{92}U^{239}} + \gamma$$

(b) Particle-particle reactions : The bombarding particle is absorbed to form the compound nucleus which disintegrates to form the product nucleus along with a massive particle like proton, neutron, etc. Such reactions are commonly encountered.

$$_7N^{14} + {_0n^1} \longrightarrow {_6C^{14}} + {_1H^1}$$

$$_9F^{19} + {_0n^1} \longrightarrow {_7N^{16}} + {_2He^4}$$

$$_5B^{11} + {_1H^1} \longrightarrow {_6C^{11}} + {_0n^1}$$

$$_7N^{14} + {_1H^1} \longrightarrow {_6C^{11}} + {_2He^4}$$

(c) Fission reactions : In these a heavy nucleus, when bombarded with particles, breaks into two or more fragments, with masses roughly half that of original along with or without the emission of light particles. This process produces large amounts of energy and usually several neutrons. For example, the fission of U^{235} by neutrons may be represented as :

$$_{92}U^{235} + {_0n^1} \longrightarrow {_{56}Ba^{141}} + {_{36}Kr^{92}} + 3\ {_0n^1}$$

The presence of barium was identified by chemical methods in the disintegration products. These days fission of thorium, protactinium and uranium isotopes with protons, neutrons, deuterons, alpha particle and gamma rays have been studied. Fission of lighter isotopes with bombarding particle is also made.

(d) Spallation reactions : Such reactions were discovered by G. T. Seaborg and J. P. Periman in 1947 at Berkeley.

Many nuclei when bombarded with high energy charged particles break up giving products of normal type and a large number upto 30 units and in charge upto 40 result. Thus it differs from nuclear fission in which products of identical masses are obtained.

When arsenic ($_{33}As^{75}$) is bombarded by high energy alpha particle ($_2He^4$) one of the minor products is chlorine of mass number 38, *i.e.*, $_{17}Cl^{38}$. The net decrease in the number of protons is thus 16. This particular mode of spallation is then expressed by :

$$_{33}As^{75} + {}_2He^4 \longrightarrow {}_{17}Cl^{38} + \text{some other particles.}$$

Another spallation reaction may be indicated by the following equation.

$$_{29}Cu^{63} + {}_2He^4 \text{ (400 Mev)} \longrightarrow {}_{17}Cl^{37} + 14\,{}_1H^1 + 16\,{}_0n^1.$$

(e) Fusion reaction. Certain light nuclei may fuse together to produce heavier nuclei, *e.g.*, in the formation of helium from hydrogen. These reactions are known as fusion reactions. These reactions are exothermic and of considerable interest as potential source of energy,

$$_1H^2 + {}_1H^2 \longrightarrow {}_2He^3 + {}_0n^1 + 3.25 \text{ MeV}$$

$$_1H^2 + {}_1H^2 \longrightarrow {}_1H^3 + {}_1H^1 + 40 \text{ MeV}$$

$$_1H^3 + {}_1H^2 \longrightarrow {}_2He^4 + {}_0n^1 + 17.8 \text{ MeV}$$

$$_1H^2 + {}_2He^3 \longrightarrow {}_2He^4 + {}_1H^1 + 18.3 \text{ MeV}$$

2. In terms of bombarding agents : Classificational nuclear reactions in terms of the bombarding agents is explained in the following ways :

(a) Alpha induced reactions : Because of large coulombic repulsions, reactions of this type are comparatively inefficient.

(α, n) **type :**

$$_4Be^9 + {}_2He^4 \longrightarrow {}_6C^{12} + {}_0n^1$$

$$_{13}Al^{27} + {}_2He^4 \longrightarrow {}_{15}Si^{30} + {}_0n^1$$

(α, p) type : $_7N^{14} + {}_2He^4 \longrightarrow {}_8O^{17} + {}_1H^1$

$$_{13}Al^{27} + {}_2He^4 \longrightarrow {}_{14}Si^{30} + {}_1H^1.$$

The reaction $_4Be^9$ (α, *n*) $_6C_{12}$ provides a convenient method for the production of neutrons.

(b) Proton-induced reactions : Although coulombic repulsions are less with protons than with the alpha particles, the smaller masses of the protons makes it difficult to surpass or penetrate the energy barrier unless it is highly energised. Proton reactions commonly yield positron emitters because of unfavourable increase in nuclear charge.

(p, γ)

$$_7N^{14} + {}_1H^1 \longrightarrow {}_8O^{15} + \gamma$$

$$_6C^{12} + {}_1H^1 \longrightarrow {}_7N^{13} + \gamma$$

(p, n) $_5B^{11} + {}_1H^1 \longrightarrow {}_6C^{11} + {}_0n^1$

(p, d) $_4Be^9 + {}_1H^1 \longrightarrow {}_4Be^8 + {}_1H^2$

$$_7N_{14} + {}_1H^1 \longrightarrow {}_6C^{11} + {}_2He^4$$

$$_{13}Al^{27} + {}_1H^1 \longrightarrow {}_{12}Mg^{24} + {}_2He^4$$

(c) Deuteron induced reactions : Deuterons are perhaps the most effective of the positively charged particles. Because of a comparatively large mass effect, the deuteron is inherently energetic. Further increase in energy may be supplied by cyclotron acceleration.

(d, α)

$$_3Li^6 + {}_1H^2 \longrightarrow {}_2He^4 + {}_2He^4$$

$$_8O^{16} + {}_1H^2 \longrightarrow {}_7N^{14} + {}_2He^4$$

(d, p)

$$_6C^{12} + {}_1H^2 \longrightarrow {}_6C^{13} + {}_1H^1$$

$$_{15}P^{11} + {}_1H^2 \longrightarrow {}_{15}P^{32} + {}_1H^1$$

In such reactions nuclear charge remains the same but neutron number is increased by one.

(d, n)

$$_6C^{12} + {}_1H^2 \longrightarrow {}_7N^{13} + {}_0n^1$$

$$_{17}Cl^{37} + {}_1H^2 \longrightarrow {}_{18}A^{38} + {}_0n^1$$

(contd...)

Certain (d, n) reactions are excellent sources.

(d, 2n)

An example of this is

$$_{17}Cl^{37} + {}_{1}H^{2} \longrightarrow {}_{18}Ar^{37} + 2\,{}_{0}n^{1}$$

(d) Gamma-induced reactions : Experimentally, it is observed that gamma-induced reactions can only take place if the energy of the gamma-ray is enough to liberate the nuclear particle from its binding inside the nucleus. The phenomenon is known as *nuclear photo effect.* Here Y-quantum may be regarded as just another kind of nuclear projectile. Some examples are cited below :

(i) *Low energy gamma ray bring about* (γ, *n*) *reactions.*

$$_{1}H^{2} + \gamma \longrightarrow {}_{1}H^{1} + {}_{0}n^{1}$$

$$_{4}Be^{9} + \gamma \longrightarrow {}_{4}Be^{8} + {}_{0}n^{1}$$

(ii) (γ, *p*) *reaction may also occur with j-rays of higher energy.*

$$_{13}Al^{27} + \gamma \longrightarrow ({}_{13}Al^{27})^{*} \longrightarrow {}_{11}Na^{24} + {}_{1}H^{1} + {}_{1}H^{1} + {}_{0}n^{1}$$

The incident particle energies are large enough so that more than one nucleon can be released from the nucleus; a spallation of nucleus takes place.

(e) Neutron induced reactions : In the absence of coulombic repulsions, neutron- induced reactions are dependent upon the energy of the bombarding particle and the cross section of the nucleus in question. The nucleus absorbs a neutron yielding the excited compound nucleus. The excited compound nucleus loses its energy in the form of either.

(i) the emission of gamma radiation (n, γ)

(ii) the ejection of an alpha panicle (n, α)

(iii) the ejection of proton (n. p) or

(iv) fission (n, f).

Nuclear Cross Section

The probability of a nuclear process is generally expressed in terms of cross section ρ which has the dimensions of an area.

The total cross section for collision with particle is never greater than the geometrical cross sectional area of the nucleus and therefore cross sections are generally not large than 10^{-24} cm^2 (radii of the heaviest nuclei are about 10^{-12} cm). Hence, a cross section of 10^{-24} has been named as **barn** *i.e.*,

$$1 \text{ barn} = 10^{-24} \text{ cm}^2$$

Millibarn (10^{-27}cm^2) and the microbarn (10^{-30} cm^2) are also used.

The total cross section σ_t, is equal to the sum of the absorption cross section σ_a (also called the reaction cross section) and scattering cross section σ_s, and fission cross section, σ_f. Thus,

$$\sigma_t = \sigma_\alpha + \sigma_s + \sigma$$

If we consider a thick target, the reaction is

$$I_0 - I = I_0 (I^{-2}\sigma^x)$$

where I is the intensity of beam after traversing a target of thickness *x*, I_0 is the incident intensity and $I_0 - I$ is the number of reactions occurring.

ARTIFICIAL TRANSMUTATION

Introduction : The nuclei of natural radioactive elements are unstable. They are disintegrating continuously emitting α, β and γ rays forming new nuclei. In 1919 Rutherford first discovered that stable nuclei could be disintegrated by bombarding them with high energy particles. *This process of disintegration of stable nuclei by artificial means is called the artificial disintegration of elements.* Artificial disintegration is also known as *artificial transmutation* since it can be used to convert one element into another.

Discovery of Artificial Transmutation. The simple apparatus used by Rutherford to demonstrate the first artificial transmutation. A radioactive substance, capable of emitting α-particle is placed on the disc *D* whose distance from the screen *S* could be varied. *E* is a silver foil whose thickness is enough to absorb the α-particles ejected from the source placed on *D*.

'*S*' is a fluorescent screen coated with zinc sulphide and is placed outside the opening, and scintillations on the screen could be observed with the help of a microscope M. When oxygen or carbon dioxide was introduced, no scintillation was produced indicating no effect of α-particles on O_2 or CO_2. When nitrogen was kept in place of O_2 or CO_2 and the distance between *D and F* was about 40 cm, scintillations are observed on the screen. These scintillations could not be due to α-particles because of their short range but were due to particles ejected from the nitrogen nucleus by impact of α-particles. An examination of these particles by cloud-chamber shows that these are protons.

$$_7N^{14} + {}_2He^4 \longrightarrow {}_8O^{17} + {}_1H^1$$

In simple words, the process involves the absorption of α-particles by the nucleus of nitrogen, which now being in a state of a tension, immediately disintegrates to give a proton and an oxygen isotope. Thus, nitrogen is transmutated to oxygen.

Later on Rutherford and Chadwick extended the work to other lighter elements and they found positive results in elements from boron to potassium with the exception of oxygen and carbon. They also found that in some cases energy carried out by protons was *greater then* the α-particles. This result provided additional proof that the protons are the results of disintegration and the extra energy was acquired as a result of nuclear rearrangements.

Some examples of these reactions are :

(i) $_5B^{10} + {}_2He^4 \longrightarrow [{}_7N^{14}]^* \longrightarrow {}_6C^{13} + {}_1H^1$

(ii) $_{19}K^{39} + {}_2He^4 \longrightarrow [{}_{21}Sc^{43}]^* \longrightarrow {}_{20}Ca^{42} + {}_1H^1$

In general, the (α, *p*) reactions may be represented as

$$_zA^{\alpha} + {}_2He^4 \longrightarrow [{}_{z+2}C^{\alpha+4}]^* \longrightarrow {}_{z+1}B^{\alpha+3} + {}_1H^1$$

The element *A* is called the target ; *C* the compound nucleus and *B* the product element.

Production of high speed particles. It is not possible to cause transmutation on large scale by α-particles obtained from natural sources. If one gram atom is to be transmuted, a process is needed which should be 10^{21} times as effective as the Rutherford process. Cockfit devised an apparatus called *linear accelerator* in which particles which are charged by strong electrostatic field are accelerated. The most important of the high speed particle generators was first devised in 1939 by Lawrence and is known as *cyclotron.* This is a powerful weapon in transmutation technique with great potentiatlities.

In cyclotron the particle is accelerated in several steps using the same source of potential each step being so synchronised that it produces an accelerating influence on the particle at each stage.

Other particles like protons, neutrons and deuterons etc. have been tried as bombarding particles. With the invention of particle accelerators, fast moving particles have been made to bombard the different types of targets.

(i) Neutrons : Fast neutrons used for bombardment of beryllium or lithium have also been used for disintegrating other nuclei. Being uncharged, they can approach the nucleus of atom without being repulsed.

Reactions between nuclei and neutrons give rise to α-paricles, protons, photons or neutrons. Some examples are :

$$_3Li^6 + {_0n^1} \longrightarrow [_3Li^7]^* \longrightarrow {_1H^3} + {_2He^4}$$

$$_{13}Al^{27} + {_0n^1} \longrightarrow [_{13}Al^{28}]^* \longrightarrow {_{13}Al^{28}} + h\nu$$

$$_{29}Cu^{63} + {_0n^1} \longrightarrow [_{29}Cu^{64}]^* \longrightarrow {_{28}Cu^{62}} + 2\ {_0n^1}$$

(ii) Protons : This was first investigated by Cockcroft and Walton in 1932. They bombarded a lithium target with protons accelerated to energies from 100 to 700 Mev in a high voltage discharge tube. The two α-particles thus produced were observed by scintillation method. The reaction is

$$_3Li^7 + {_1H^1} \longrightarrow {_2He^4} + {_2He^4}$$

Later, protons bring about artificial disintegration, emitting not only α-particles but deuterons, neutrons and γ-rays also.

$$_4Be^9 + {_1H^2} \longrightarrow {_4Be^8} + {_1H^3}$$

$$_{20}Ca^{44} + {_1H^1} \longrightarrow {_{21}Sc^{44}} + {_0n^1}$$

(iii) Deuterons : Many nuclear disintegrations have been performed by using deuterons, accelerated to energies upto several hundred million electron volts in a cyclotron or synchrocyclotron. In these reactions, α-particles, protons, neutrons and tritium are produced. For instance,

$$_3Li^6 + {_1H^2} \longrightarrow {_2He^4} + {_2He^4}$$

$$_6C^{12} + {_1H^2} \longrightarrow {_6C^{13}} + {_1H^1}$$

Deuteron can also interact with deuteron in a manner which produces a special isotope of helium and a neutron.

$$_1D^2 + {_1D^2} \longrightarrow {_2He^3} + {_0n^1}$$

(iv) Gamma rays : γ-rays can bring about artificial disintegration provided its energy is greater than the binding energy of the nucleus, y-rays may be obtained from natural radioactive sources or produced by the impact of electrons accelerated to high energies by Van de Graaff generator. Some reactions are :

$$_1H^1 + _0n^1 \quad _1H^2 + h\nu$$

$$_4Be^9 + h\nu \longrightarrow _4Be^8 + _0n^1$$

Detection and measurement of artificial transmutations :

Various techniques have been developed for the detection and measurement of artificial transmutation. Some of these are like ;

1. Scintillation method
2. Wilson cloud chamber method.
3. Geiger Muller counter method.
4. Magnetiç spectrograph method.

ARTIFICIAL RADIOACTIVITY

Several of the known isotopes are man made. They are usually unstable. When elements are bombarded with fast moving projectiles, the products are often radioactive. They disintegrate in a definite time which is characteristic of every product.

Artificial radioactivity was discovered in 1934 by Irene Curie and F. Joliot when bombarded boron, the emission of positron ($_{-1}e^0$, positron has the same mass as electron but carries positive charge), proton and neutron.

The emission of protons and neutrons was stopped as soon as the bombarding source was removed but not of positrons. Evidently in this phenomenon an unstable isotope is initially produced which decays to a stable isotope by positron emission. *This process of converting of a stable element into radioactive element is called artificial or induced radioactivity.* In the above examples $_7N^{*13}$ and $_{15}P^{*30}$ are artificially produced ratio elements.

Artificial radioactivity by different bombarding particles :

(a) By alpha ray bombardment Radioactive nuclides are produced in both (α, p) and (α, n) reactions. In the (α, p) process usually stable isotopes are produced but sometimes unstable nuclei are also produced which themselves emit electrons. In the (α, n) process, the nuclei produced always emit positrons.

(i) $_{12}Mg^{25} + {_2He^4} \longrightarrow {_{13}Al^{28}} + {_1H^1}$

$$_{13}Al^{28} \longrightarrow {_{14}Si^{28}} + {_{-1}e^0}$$

(ii) $_{13}Al^{27} + {_2He^4} \longrightarrow {_{15}P^{30}} + {_0n^1}$

$$_{15}P^{30} \longrightarrow {_{14}Si^{30}} + {_{+1}e^0}$$

(iii) $_4B^{10} + {_2He^4} \longrightarrow {_7N^{13}} + {_0n^1}$

$$_7N^{13} \longrightarrow {_6C^{13}} + {_{+1}e^0}$$

(b) By deuteron bombardment : The (d, p) and (d, α) reactions very often yield radioactive species. The examples are :

(i) $_{11}Na^{23} + {_1D^2} \longrightarrow {_1H^1} + {_{11}Na^{24}}$

$$_{11}Na^{24} \longrightarrow {_{12}Mg^{24}} + {_{-1}e^0}$$

(ii) $_{16}S^{32} + {_1D^2} \longrightarrow {_{15}P^{30}} + {_2He^4}$

$$_{15}P^{30} \longrightarrow {_{14}Si^{30}} + {_{+1}e^0}$$

(iii) $_{12}Mg^{26} + {_1D^2} \longrightarrow {_{11}Na^{24}} + {_2He^4}$

$$_{11}Na^{24} \longrightarrow {_{12}Mg^{24}} + {_{-1}e^0}$$

The (d, n) reactions also yield unstable isotopes.

(iv) $_8O^{16} + {_1D^2} \longrightarrow {_9F^{17}} + {_0n^1}$

$$_9F^{17} \longrightarrow {_8O^{17}} + {_{+1}e^0}$$

An interesting example of deuteron bombardment is that of bismuth. The half-life of the product of deuteron bombardment of bismuth is five days and this emits β-rays and ultimately an α-ray emitter (polonium) is produced.

It can therefore be concluded that the first unstable element was radium-E and since only one stable isotope of bismuth (Bi^{209}) is known, the disintegration equation would be as follows

$$_{83}Bi^{209} + {_1D^2} \longrightarrow {_{83}\text{Ra–E}^{210}} + {_1H^1}$$

(c) By gamma rays : Induced or artificial radioactivity is also produced by the action of gamma rays :

$$_{15}P^{31} + \gamma \longrightarrow {_{15}P^{30}} + {_0n^1}$$

$$_{15}P^{30} \longrightarrow {}_{14}Si^{30} + {}_{+1}e^{0}$$

(d) By neutron bombardment : When a number of elements are bombarded with neutrons, radio-elements are produced by (n, α), (n, p) and (n, γ) reactions. The various examples are :

(i) (n, a) reaction.

$$_{13}Al^{27} + {}_{0}n^{1} \longrightarrow {}_{2}He^{4} + {}_{11}Na^{24}$$

$$_{11}Na^{24} \longrightarrow {}_{12}Mg^{24} + {}_{-1}e^{0}$$

(ii) (n, p) reaction.

$$_{11}Na^{23} + {}_{0}n^{1} \longrightarrow {}_{18}Ne^{23} + {}_{1}H^{1}$$

$$_{10}Na^{23} \longrightarrow {}_{11}Ne^{23} + {}_{-1}e^{0}$$

(iii) **(n, γ) reaction.**

$$_{18}Ar^{40} + {}_{0}n^{1} \longrightarrow {}_{18}Ar^{41} + \gamma$$

$$_{18}Ar^{41} \longrightarrow {}_{19}K^{41} + {}_{-1}e^{0}$$

(iv) (n, 2n) reaction.

$$_{19}K^{39} + {}_{0}n^{1} \longrightarrow {}_{19}K^{38} + 2\ {}_{0}n^{1}$$

$$_{19}K^{38} \longrightarrow {}_{18}Ar^{38} + {}_{+1}e^{0}$$

The most important results in the artificial production of radioactive isotopes have been achieved by neutrons bombardment. Fermi discovered that most of the elements, when bombarded with neutrons, slowed down by passage through water or paraffin wax, gave radioactive isotopes.

Mechanism of Artificial Radioactivity. The mechanism of the artificial radioactivity is as follows:

(i) *When nuclei of lighter elements are bombarded by α-particles, protons or neutrons are thrown out of the nucleus resulting in an* ***unstable or disturbed*** *nucleus which on returning to stable state emits out radioactive radiations.*

(ii) *In addition to the emission of α, β-particles and γ-radiation, emission of positrons and capture of orbital electrons take place during the artificial disintegrations.*

The orbital electron capture is known as K–electron capture.

Explanation of Emission of Positrons ($_{+1e}{}^{0}$). The emission of the positron is explained by the transformation of a proton into a neutron and positron inside the nucleus

$$p \longrightarrow n + {}_{+1}e^{0}$$

The stability of lighter nuclei is governed by the proton to neutron ratio. The emission of positron alters this ratio in favour of a more stable nucleus.

On the other hand if an electron is emitted, the neutron-proton ratio (N/P) is lowered and the nucleus tends to become stable.

Explanation of K-electron capture. The unstable nucleus captures an electron from the nearest energy shell, *i.e., K* shell. This is followed by the fall of an electron from a higher shell to the *'K'* shell to fill the vacancy caused by the captured electron. This results finally in the release of energy in the form of radiations.

The K–capture takes place in preference to positron emission because in positron emission neutron is also emitted.

An example of K-electron capture is given below :

$${}_{25}Mn^{54} + {}_{-1}e^{0}((K) \longrightarrow {}_{24}Cr^{54} + \gamma$$

(iii) The nuclei with extremely small number of protons attain stability by increasing their nuclear charge while nuclei with very large number of protons attain stability by decreasing its nuclear charge. The emission of electron increases the charge whereas the emission of positron decreases charge. Therefore, usually isotope of an element with small mass number than the stable isotopes emits electrons and those with mass number greater than those of the stable isotopes emit positrons.

According to wave mechanics, extra nuclear electrons during their motions, often approach very close to the nucleus and sometimes penetrate the nucleus. The electrons of K-shell do so in elements with high atomic number. The process is known as K-capture. The electron may also jump from L-shell but it is less probable. These K–and L–electrons are compensated by outer electrons giving rise to *K* or *L,* X–ray spectra, respectively and no charged particle is emitted. Thus, in all three isobaric transformation occurs because nuclear charge changes but mass number remains the same. Sometimes orbital electron capture results into emission of electrons. These electrons are known as *Auger electrons.* Just to know whether the disintegration involves electron emission, positron emission or electron capture, one has to examine the energy terms available for each type of disintegration.

(a) When a positron is emitted, then

$$_{Z}X^{A} \longrightarrow {}_{Z-1}Y^{A} + {}_{+1}e^{0}$$

The energy balance for the above type of disintegration may be expressed as follows :

$$\frac{E}{C^2} = m_n\ ({}_{Z}X^{A}) - mn\ ({}_{Z-1}Y^{A}) - m \qquad ...(1)$$

where m_n = the nuclear masses of artificial nuclide and its decay product, m = mass of positron, E = energy and C = the velocity of light. The equation (1) may be put as follows in terms of atomic masses

$$\frac{E}{C^2} = m_a\ ({}_{Z}X^{A}) - Z.m - m_a\ ({}_{Z-1}Y^{A}) + (Z-1)\ m - m \qquad ...(2)$$

where m_a = atomic mass.

or $$\frac{E}{C^2} = m_a({}_{Z}X^{A}) - m_a\ ({}_{Z-1}Y^{A}) - 2_m \qquad ...(3)$$

For a positron emission, E must be positive and for that it is necessary *that the atomic mass of the artificial nuclide must be greater than the atomic mass of its isobar with nuclear charge one unit smaller.* Thus,

$$m_a({}_{Z}X^{A}) > m_a\ ({}_{Z-1}Y^{A}) + 2m \qquad ...(4)$$

The nuclei which usually decay with positron emission are C^{11}, N^{13}, O^{15}, F^{16}, Na^{21}, Mg^{23}, and Al^{25},

When electron emission occurs, then

$$_{Z}X^{A} \longrightarrow {}_{Z+1}Y^{A} + {}_{-1}e^{0}$$

For this,

$$E/C^2 = m_n({}_{Z}X^{A}) - m_n({}_{Z+1}Y^{A}) - m \qquad ... (5)$$

or $$E/C^2 = m_n({}_{Z}X^{A}) - Z.m - m_a({}_{Z+1}Y^{A}) + (Z+1)\ m - m \qquad ...(6)$$

$$= m_a({}_{Z}X^{A}) - m_a({}_{Z+1}Y^{A}) \qquad ...(7)$$

As the value of E should be positive for an electron emission, it means that,

$$m_a({}_{Z}X^{A}) > m_a({}_{Z+1}Y^{A})$$

Hence it may be concluded that *the atomic mass of artificial nuclide should be greater than that of its isobar with nuclear charge one unit greater.*

Nuclides which decay with electron emission are C^{14}, N^{16}, Ne^{23}, Na^{24}, Mg^{27}, Al^{28} etc.

When an electron is captured then

$$_{Z}X^{A} + {}_{-1}e^{0} \longrightarrow {}_{Z-1}Y^{A}$$

For this equation, is

$$E/C^2 = m_n({}_{Z}X^{A}) + m - m_n({}_{Z-1}Y^{A}) \quad \text{... (8)}$$

$$= m_a({}_{Z}X^{A}) - Z.m + m + m_a({}_{Z-1}Y^{A}) + (Z-1)\,m \quad \text{...(9)}$$

$$E/C^2 = m_a({}_{Z}X^{A}) - m_a({}_{Z-1}Y^{A}) \quad \text{...(10)}$$

Hence for electron capture

$$m_a({}_{Z}X^{A}) > m_a({}_{Z-1}Y^{A}) \quad \text{...(11)}$$

Therefore, for electron capture the *atomic mass of the artificial nuclide must be greater than its isobar with nuclear charge one unit smaller.*

From the comparison of equations (4) and (11) we conclude that electron capture is more probable than positron emission, K–electron capture has been found to depend on the small probability of electron being too close or being within the nucleus. As soon as the energy exceeds the value $2mC^2$ positron emission takes place more favourably than the electron capture. The electron capture is favourable only when equation (11) is satisfied, which is rarely the case.

Applications

(i) *This has been used in the synthesis of elements after atomic number* 92.

(ii) *Artificial radioactivity has been" used to prepare radioactive isotopes which find wide uses in medicine, agriculture and industry.*

(iii) *It provides a sensitive method for study of complex phenomena such as photo-synthesis.*

Conclusion : The fundamental difference in the break-down of artificial and natural radioactivity is that in the latter case the products are electrons or α-particles, while in the former case the products may be electrons or positrons. It is possible to predict in general, which particle will be emitted, if the radioelement has a smaller proton/neutron ratio in its nucleus than the corresponding stable isotope, then an electron will be emitted. On the other hand if the proton/neutron ratio is too high, positron particle will be emitted.

RUTHERFORD AND SODDY'S THEORY OF RADIOACTIVE DISINTEGRATION

In 1902, **Rutherford** and **Soddy** formulated a theory to explain the spontaneous disintegration of radioactive elements.

(i) *Atoms of every radioactive element are constantly breaking up into fresh radio-active products with the emission of* α *and* β *and* γ *rays.*

(ii) *The rate of disintegration is not influenced by external factors such as temperature,* pressure, chemical combination, etc. *but is entirely dependent upon the law of chance, i.e., the number of atoms breaking per second at any instant is proportional to the number present, at that instant, i.e.,*

$$-\frac{dN}{dt} \propto N \qquad \text{...(1)}$$

where dN represents the number of nuclei which decay during the time interval out of N nuclei present at time t. As the rate of decay is independent of pressure and temperature, this implies that the activation energy of radioactive decay is zero.

In equation (1), a minus sign has been introduced to take into account of the fact that dN represents a decrease in the number of nuclei present during the positive time interval dt.

Equation (1) can be written as

$$\frac{dN}{dt} = -\lambda N \qquad \text{...(2)}$$

where λ is called the radioactive constant and is a definite and specific property of a given radioelement. Equation (2) can be written as

$$\frac{dN}{N} = -\lambda Nt$$

Integrating it, we get

$$\log_e N = -\lambda t + C \quad ...(3)$$

where C is a constant of integration.

Now when $t = 0$, $N = N_0$; equation (3) becomes as

or $$C = \text{loge } N_0 \quad ...(4)$$

On combining equations (3) and (4), we get

$$\text{loge } N = -\lambda t + \text{loge } N_0$$

or $$\text{loge } \frac{N}{N_0} = -\lambda t \text{ or } N/N_0 = e^{-\lambda t} \quad ...(5)$$

From equation (5), it follows that the number of neuclei in radioactive element decreases exponentially with time.

Activity. The activity 'A' of a radioactive substance is the rate of decay, *i.e.*, number of disintegrations per second.

$$A = -\frac{dN}{dt} = \lambda N \quad ...(6)$$

Also, the activity, at time $t = 0$,

$$A_0 = -\frac{dN_0}{dt} = \lambda N_0 \quad ...(7)$$

where N_0 is the number of nuclei present at $t = 0$. Dividing equation (6) by (7), we get

$$\frac{A}{A_0} = \frac{N}{N_0}$$

But $$\frac{N}{N_0} = e^{-\lambda t} \quad \text{[from equation (5)]}$$

$$\therefore \quad \frac{A}{A_0} = e^{-\lambda t} \quad ...(8)$$

The unit of activity is the curie which is defined as that quantity of radioactive material which decays at the rate of 3.70×10^{10} disintegrations per second.

Half-Life Period. According to the exponential law, an infinite time is required theoretically to disintegrate a radioactive element completely. Hence a quantity known as half-life period is commonly used. *It is defined as that tie in which* half of the initial radioactive atoms are disintegrated.

If N_0 nuclei are present initially and N_1 are the number of atoms present at a time t_1, then

$$N_1 = N_0 e^{-\lambda t_1}$$

When $t_1 = T_{1/2}$, $N = \frac{N_0}{2}$

$$\frac{1}{2} N_0 = N_0 e^{-\lambda T_{1/2}} \text{ or } e^{\lambda T_{1/2}} = 2$$

$$\lambda T_{1/2} = \log_e 2$$

Half-life period. $T^{1/2} = \frac{\log_e 2}{\lambda} = \frac{0.693}{\lambda}$

Alternative Method :

If N_1 nuclei are present at time t_1 and one half of the number, $N_2 = N_1/2$ have survived at time t_2, we can write

$$N_1 = N_0 e^{-\lambda t_1} \quad ...(9)$$

$$N_2 = N_0 e^{-\lambda t_2} \quad ...(10)$$

Dividing equation (9) by (10), we get

$$\frac{N_1}{N_2} = e^{\lambda (t_2 + t_1)}$$

$$N_2 = \frac{N_1}{2} \text{ But}$$

$$\frac{N_1}{N_1/2} = e^{\lambda (t_2 + t_1)} \therefore$$

$$2 = e^{\lambda (t_2 + t_1)}$$

Taking logarithm, we get

$$\log_e 2 = \lambda (t_2 - t_1)$$

or $$2.303 \log_{10} 2 = \lambda (t_2 - t_1)$$

or $$2.303 \times C.3010 = \lambda T_{1/2} \quad [\because \log_{10} 2 = 0.3010]$$

or $$T_{1/2} = \frac{0.693}{\lambda} \quad ...(11)$$

The half-lives of radioactive nuclides very considerably for different elements, from 10^{15} years for the longest lived to 10^{-11} sec for the shortest lived known nuclide.

Determination of $T_{1/2}$. The value of the decay constant λ can be determined experimentally from which $T_{1/2}$ can be evaluated. In the equation log N = log N_0 – lt, the counting rate N is observed at known intervals t and log (N/N_0) is plotted against t giving λ as the slope. Half lives of radioactive substance vary to a great deal. Polonium-212 is an α-emitter and has a half life of 3×10^{-7} sec, whereas thorium for instance, has a half-life of 1.391×10^{11} years and is almost stable.

If the half-life of a radioactive substance is either very short or very long, then the method outlined above is unsuitable. If the half-life is very long then l is quite small, it may not be possible to detect a change in *activity* during the course of measurements. The experimental activity A is defined equal to CλN where C is the *detection coefficient* which depends upon the nature and efficiency of the detection instrument. λN is equal to the fraction of the disintegrating atoms detected by the measuring device as given by.

The value of C is obtained through calibration.

This method is successful for long lived elements having half-life of the order of 10^{10} years. Short half-lives are determined by somewhat complex techniques.

Mean or Average Life. *Average life of a radioactive substance is defined as the ratio of the total life of all the radioactive atoms to the total number of such atoms in it.* In other words average life of a radioactive substance is when all the radioactive substance is disintegrated, *i.e.*, the value of $N_1 = 0$ in the exponential law.

$$N_1 = N_0 e^{-\lambda t}$$

or
$$\log_e \frac{N_1}{N_0} = -\lambda t$$

or
$$\text{loge}\left(1 - \frac{N_0 - N_1}{N_0}\right) = -\lambda t$$

Expanding the above logarithmic term and neglecting higher powers of $\left(\frac{N_0 - N_1}{N_0}\right)$, we get

$$-\left(\frac{N_0 - N_1}{N_0}\right) = -\lambda t$$

or $$\frac{N_0 - N_1}{N_0} = -\lambda t$$

When N1 = 0, t = TA (average or mean life) $= \frac{1}{\lambda}$

which is independent of concentration term and characteristic of the disintegrating element.

Alternative Method. When a radioactive substance is separated from its disintegration products, the nuclei which disintegrate earlier have a very short life and others which disintegrate at the end have a long life. Hence it is useful to determine the mean life or average life of the radioactive substance which is defined as :

$$\text{Average Life} = \frac{\text{Sum of lives of all the nuclei}}{\text{Total No. of nuclei}}$$

If dN is a small group of nuclei which decay during a time interval between t and t + dt, the combined life-time of this is : dT = tdN. Adding all these life-times for the group of nuclei that decay during the entire time interval from t = 0 to t = ∞, we get for the total combined life of all the nuclei.

$$T = \int_0^\infty dT = \int_0^\infty tdN$$

Dividing this by all nuclei N_0, initially present, we get mean or average life T_A.

$$T_A = \frac{1}{N_0} \int_0^\infty tdN$$

$$= \frac{1}{N_0} \int_0^\infty td(N_0 e^{-\lambda t}) \quad [\because N = N_0 e^{-\lambda t}]$$

$$= \frac{1}{N_0} \int_0^\infty t.N_0 - \lambda.e^{-\lambda t}$$

$$= \lambda \int_0^\infty te^{-\lambda t}\, dt$$

(– ve sign has been omitted since it only represents the process of disintegration)

$$T_A = \lambda \left[\frac{t.e^{-\lambda t}}{-\lambda}\right] - \lambda \int_0^\infty \frac{e^{\lambda t}}{-\lambda} \cdot dt$$

$$= [-t.e^{-\lambda t}]_0^{\infty} + \int_0^{\infty} e^{-\lambda t}.dt$$

$$= 0 + \int_0^{\infty} e^{-\lambda t}.dt$$

$$= \left[\frac{e^{-\lambda t}}{-\lambda}\right]_0^{\infty} = \frac{1}{\lambda}$$

Hence the *average life of a radioactive element is reciprocal of the radioactive constant.* The average half-lives of the natural radio elements very form 10^{-9} sec to 10^{10} years so that the range is enormous.

Radioactive constant. Radioactive constant 'λ' is a definite and specific property of a given radio element. Its value depends only on the nature of radioactive element and is independent of the physical condition and state of chemical combination.

We know

$$N = N_0\, e^{-\lambda t} \quad ...(13)$$

In time $t = \frac{1}{\lambda}$, equation (13) becomes as

$$\frac{N}{N_0} = e^{-\lambda t}.\frac{1}{\lambda} = e^{-1} = \frac{1}{e} \quad ...(14)$$

Hence, the radioactive constant is defined as *"the reciprocal of the time during which the number of radioactive nuclei falls to 1/e of its original value.* The decay constant λ has the dimensions of sec^{-1}.

Activity of a mixture. If a sample consists of a mixture of independently decaying radio-isotopes (such as phosphorus-32 and sodium-24), the presence of one will have no effect on the decay rate of the other, and the total activity of the sample will merely be the sum of the individual activities.

Thus, the total activity *'A'* of the mixture is given by the equation.

$$A = A_1 + A_2 + ...$$

or $$A = \left(-\frac{dN_1}{dt}\right) + \left(-\frac{dN_2}{dt}\right) + ...$$

or $$A = \lambda_1 N_1 + \lambda_2 N_2 + ... \qquad \left[\because -\frac{dN}{dt} = n\lambda\right]$$

or $A = \lambda_1 N_0 e^{-\lambda_1 t} + \lambda_2 N_0 e^{-\lambda_2 t} + \ldots \quad [\because N = N0\ e^{-\lambda t}]$ ...(15)

For a simple source consisting of a single species of nuclide, the logarithm of the activity plotted against time results in a straight line plot. This is illustrated by the dotted line. If these two nuclides are mixed together, they will decay as they did separately, but the observed activity of the mixture (the sum of the two activities) is given by the solid curved line.

Radioactive equilibria. It is seen that in many cases the product nuclei from radioactive decay are themselves radioactive. If the product (or daughter nucleus) shows a characteristic activity of its own, the resultant activity of a sample of the material will, in general be more complex than if there were only a single type of radioactive nuclide present in the sample.

Consider a radioactive element *'A'* which disintegrates to yield a daughter element *'B'*. If this daughter element is radioactive, it will, in turn, disintegrate to form another element *C*.

$$A \longrightarrow B \longrightarrow C$$

Let N_1 be the amount of parent substance *'A'* at any instant and λ_1 its decay constant, while N_2 be the amount of daughter *'B'* at the same instant. Let N_1^0 and N_2^0 be the amount of the parent and the daughter substances respectively at $t = 0$.

$$-\frac{dN_1}{dt} = l1N1$$

and $N_1 = N_1^0\ e^{\lambda_1 t}$ ...(16)

Daughter B is formed at the rate at which A decays, *i.e.*, $\lambda_1 N_1$, and itself decays at the rate $\lambda_2 N_2$.

Thus, the net rate of increase of daughter, $\frac{dN_2}{dt}$ is

$$\frac{dN_2}{dt} = \lambda_1 N_1 - \lambda_2 N_2 \qquad ...(17)$$

$$= \lambda_1 N_1^0 e^{-\lambda_1 t} - \lambda_2 N_2 \qquad ...(18)$$

or $$\frac{dN_2}{dt} + \lambda_2 N_2 - \lambda_1 N_1^0 e^{-\lambda_1 t} = 0$$

Solution of this linear differential equation for N_2 as a function of time t is

$$N_2 = \frac{\lambda_1}{\lambda_2 - \lambda_1} N_1^0 \left(e^{-\lambda_1 t} - e^{-\lambda_2 t}\right) + N_2^0 e^{-\lambda_2 t} \qquad ...(19)$$

From the special cases where only atoms of *A* are present initially, *i.e.*, N_2^0 at $t = 0$, the last term of equation (19) is equal to zero and

$$N_2 = \frac{\lambda_1}{\lambda_2 - \lambda_1} N_1^0 \ (e^{-llt} - e^{-llt}) \qquad ...(20)$$

Let us now study some important cases;

Case I : *Secular Equilibrium* : Secular equilibrium is a limiting case of a radioactive equilibrium *in which the half-life of the parent is many times greater than the half-life of the daughter i.e.,* $\lambda_1 << \lambda_2$. The difference between the half-lives of the parent and daughter is usually a factor of 10^4 or greater, so that the activity of the parent shows no appreciable change during many half- life periods of the daughter. The decay of radium-226 to radon-222 is a typical example.

$$Ra^{226} \xrightarrow[t_{1/2} = 1662 \text{ years}]{\lambda_1 = 1.3 \times 10^{-11} \sec^{-1}} Rn^{222} \xrightarrow[t_{1/2} = 38 \text{ days}]{\lambda_2 = 2.1 \times 10^{-3} \sec^{-1}} Po^{216}$$

For this secular equilibrium between radium and radon, equation (20) can be simplified still further because λ_1 is negligible compared to λ_2. Also after a period of time *t*, *t* λ_2 becomes very great and $e^{-\lambda_2 t}$ approaches zero. Hence :

$$N_2 = \frac{\lambda_1}{\lambda_2} N_1^0 \, e^{-\lambda_1 t} \qquad ...(21)$$

and substitution of equation $N_1 = N_1^0 e^{-\lambda_1 t}$ gives

$$N_2 = \frac{\lambda_1}{\lambda_2} . N_1 \text{ or } N_1\lambda_1 = N_2\lambda_2 \qquad ...(22)$$

Equation (22) illustrates the fact that the relative number of atoms of parent and daughter are inversely proportional to their decay constants. If the second, third, etc. daughter nuclei are also radioactive, the condition for secular equilibrium is then as follows :

$$\lambda_i N_i = \lambda_{i-1} Ni - 1 = \text{constant}$$

Some well known examples are :

(i) *The build up of* 3.8 *days* Ra *from an initially pure sample of extremely long-lived parent* Ra^{226} *and*

(ii) *The build up of 5 days* **RaE** *from the* **RaD** *with a half-life of 22 years.*

Case II : *Transient Equilibrium* : Transient equilibrium is similar to secular equilibrium in that the half-life of the parent is greater than the half-life of the daughter but differs from secular equilibrium in that the half-live differs only by a small factor (about 10) rather than a large factor (10^4 or greater), *i.e.*, $\lambda_1 < \lambda_2$.

As becomes very large, $e^{-\lambda_2 t}$ becomes negligible compared to $e^{-\lambda_1 t}$, and the term $e^{-\lambda_2 t}$ approaches zero. Accordingly, equation (20) simplifies to,

$$N_2 = \frac{\lambda_1}{\lambda_2 - \lambda_1} N_1^0 e^{\lambda_1 t}$$

$$= \frac{\lambda_1}{\lambda_2 - \lambda_1} N_1 \qquad (\because N_1 = N^0{}_1 e^{-\lambda_1 t}) \quad ...(23)$$

Thus in equilibrium the ratio of two activities will be

$$\frac{A_1}{A_2} = \frac{\lambda_1 N_1}{\lambda_2 N_2} = \frac{\lambda_2 - \lambda_1}{\lambda_2} = \frac{T_1 - T_1}{T_2} \quad \text{(for large t)}$$

Thus, A_1/A_2 may have any value between 0 and 1 depending upon the ratio of l1 to λ_2. The build up of 2.2 days Ra^{223} from 18.3 days Th^{227} by a-transition is the well known example of it.

The case of many Successive Decays. It has been considered by H. Bateman who has developed a general solution to the problem. The case of many successive decays is represented by any radioactive series in which radioactive daughters are produced.

$$A \xrightarrow{\lambda_1} B \xrightarrow{\lambda_2} C \xrightarrow{\lambda_3} D \xrightarrow{\lambda_4} E \xrightarrow{\lambda_5} F \xrightarrow{\lambda_6}$$

Thus, starting with pure A, the total number of atoms N_n and of daughter at time t can be calculated.

$$N_n = C_1 e^{-\lambda_1 t} + C_2 e^{-\lambda_2 t} + ... + C_n e^{-\lambda_n t}$$

where

$$C_1 = \frac{\lambda_1 \lambda_2 ... \lambda_{n-1}}{(\lambda_2 - \lambda_1)(\lambda_3 - \lambda_1)...(\lambda_n - \lambda_1)} N_1^0$$

$$C_2 = \frac{\lambda_1 \lambda_2 ... \lambda_{n-1}}{(\lambda_1 - \lambda_2)(\lambda_3 - \lambda_2)...(\lambda_n - \lambda_2)} N_1^0$$

In the use of Bateman solution, a term must be included for each member of the radioactive series.

GROUP DISPLACEMENT LAW

As a result of an alpha ray change, the new element produced has atomic number less by two than the parent element. Thus, it will lie two groups to the left in the periodic table from the parent element.

A β-ray. (β^-) change results in an increase in atomic number by one, thus the new element will shift one group to the right of the parent element in the periodic table.

The combined effect of the alpha and beta ray changes on the element is summed up in the form of *group displacement law.* First enunciated by Soddy, Russel and Fazans in the year 1913. The law states that :

Law I : *In an alpha ray change, the new element produced has mass number less by four units and atomic number by two units and falls two columns to the left of the parent element in the periodic table.*

Law II : *In a β-ray change, the new element produced has the same mass number but atomic number increases by one than the parent element. Thus, the new element falls one column to the right of the parent element in the periodic table.*

Now-a-days these laws may be stated as follows;

1. The algebraic sum of the electric charges before the disintegration must be equal to total charges after disintegration.
2. The sum of the mass numbers of the initial particles must be equal to the sum of the mass numbers of the final particles.

These laws are illustrated with the following examples :

(i) *Polonium, atomic number* 84 *and atomic weight* 215 *lies in the group VIA of the periodic table. As a result of an α-ray change it becomes lead* (*At. No.* 82; *mass number* 211) *which belongs to group IVA of the periodic table.*

Thus lead lies two places to the left of the parent element polonium.

$$\underset{\text{VIA group}}{{}_{84}Po^{215}} \longrightarrow \underset{\text{IV group}}{{}_{82}Pb^{211}} + 2He^{4}$$

(ii) *Radioactive lead produced in the first step emits a β-particle and changes into radioactive bismuth* ${}_{83}Bi^{211}$ *which lies in the group VA, i.e., one place to the right of the parent lead.*

$${}_{82}Pb^{211} \longrightarrow {}_{83}Bi^{211} + \beta^-$$

(iii) *Radioactive bismuth* ($_{83}Bi^{211}$) *further loses another* β *particle and changes into polonium* ($_{84}Po^{211}$) *which lies in the group VIA, i.e., one place to the right of* $_{83}Bi^{211}$.

$$_{83}Bi^{211} \longrightarrow {}_{84}Po^{211} + \beta^-$$

The above three changes can be represented as :

IIIA	*IVA*	*VA*	*VIA*	*VIIA*
	$_{82}Pb^{211} \leftarrow$	———	——— $_{84}Po^{215}$	
|	β	β		
	$\longrightarrow$	—$_{83}Bi^{82}$—	$\longrightarrow {}_{84}Po^{211}$	

Units of Radioactivity : The following units are mainly employed in connection with radioactivity and biological effects of ionising radiations.

1. Curie : A familiar unit of radioactivity is the curie. Originally the term curie was defined as :

"The amount of radon in equilibrium with 1 *gm of radium"*

Later, curie was defined as :

"The quantity of any radioactive substance which undergoes the same number of disintegrations per second as one gram of pure radium."

With this definition the value of curie varied with successive refinements in the measurement of decay constant or atomic weight of radium. In 1960, a joint commission of the International Union of Pure and Applied Chemistry adopted the following definition :

"Curie is defined as that quantity of any radioactive material which disintegrates at a rate of exactly 3.7×10^{10} *per second"*

The curie is a large unit of activity and the submultiple, millicurie is frequently used :

1 curie, $C_i = 10^3$ millicurie $= 10^6$ micro curie.

1 curie, $C_i = 3.7 \times 10^{10}$ disintegrations/sec.

1 millicurie, $\mu C_i = 3.7 \times 10^7$ disintegrations/sec.

1 microcurie, ($\mu C_i = 3.7 \times 10^4$ disintegrations/sec.

2. Rutherford : Another unit is the Rutherford which is defined as :

"That amount of a radioactive substance which undergoes 10^6 *disintegrations per second." Out of these units curie is generally used.*

ISOTOPIC DILUTION METHOD

Introduction : It is an example of radiometric method of trace analysis. It is used to determine the quantity of constituent (radioactive or non-radioactive) in a mixture of closely related compounds which are difficult to separate and to estimate quantitatively by the usual conventional methods.

Discovery : Isotopic dilution analysis was introduced by Hevesy and Hofer in Denmark in 1934. However, the first systematic presentation of isotopic dilution as a method was made by Rittenberg and Foster in 1940. Subsequently the usefulness of the isotope dilution method has been reported frequently.

Type : There are three general types of isotope dilution methods. These are :

(a) Direct Isotope Dilution Method.

(b) Inverse Isotope Dilution Method.

(c) Double Dilution Method.

The above three methods are based on the same fundamental principle but they differ in technique and procedure and are applied under different circumstances.

Let us discuss these one by one.

(a) Direct isotope Dilution Method : This method is used to determine the quantity of non-radioactive substance in a mixture of closely related substances which are difficult to separate quantitatively by usual conventional methods.

This method involves the following steps :

(i) Suppose it is required to determine the mass (m) of a given non-radioactive compound in a mixture of other substances. To if add a quantity m′ of the same but isotopically labelled compound (A) having a specific activity S′. The two are then thoroughly mixed to obtain a uniform distribution.

Then make the suitable treatment of the mixture to isolate the compound in pure form. It is essential that the isolated compound be pure, but it is not necessary that all the compound is recovered from the mixture. Thus, it is possible to avoid tedious processes required for quantitative separations and frequently to carry out analyses otherwise impractical.

(iii) Now determine the specific activity of the pure isolated compound. Let it be S.

Since the total activity remains constant, the balance sheet shows the following :

Table 2

Weight	*Specific activity*
m gm of compound (*A*)	0
m′ gm of labelled compound (*A*)	S″
(*m* + *m′*) gm of mixture	S

or $(m + m')S = m'S'$

or $(m + m') = \frac{m'S'}{S}$

or $m = m'\left[\frac{S'}{S} - 1\right]$ ...(1)

When S′ > > S, equation (1) becomes as

$$m = \frac{m'S'}{S} \quad ...(2)$$

In order to determine m from equation (1) or (2), it is essential to know the two specific activities (*S* and *S″*) and the mass of added labelled compound (m′).

(b) Inverse Isotope Dilution Method : As the name implies, inverse isotope dilution is essentially the reverse of direct isotope dilution. It is only applicable when a system contains an unknown amount of an isotopically labelled substance, *i.e.,* a radioactive substance of specific activity S. In order to determine *m′* gm of a radioactive substance of specific activity S′ in a simple, *m* gm of a non-radioactive material of the same substance is added. The two are thoroughly mixed to obtain a uniform distribution. A small quantity of the substance is then removed from the system, purified and its specific activity is measured. Let it be *S.* The balance sheet for this case is as follows :

Weight	*Specific activity*
m′ gm of labelled compound	S′
m gm of un!abelled compound	0
(*m* + *m′*) gm of mixture	0

As the total activity remains constant, it means

$$(m + m')\,S = m'S'$$

or $$m'\,S' = mS + m'S$$

or $$m'\,(S' - S) = mS$$

or $$m' = m\left(\frac{S}{S' - S}\right) \quad ...(3)$$

The purity of the isolated compound is of utmost importance, particularly if more than one isotopically labelled compound is present.

(c) Double Isotope Dilution Method : This technique has been used particularly in biochemistry where simple quantitative chemical separation methods are frequently not available.

This is the extension of isotopic dilution method made by Block and Anker in 1648. This does not require a knowledge of specific activity S′ of the unknown substance.

In this technique, two aliquots of the sample are removed and different amounts of carrier *m′* and *m* are added to each. A small quantity of the substance to be determined is extracted from each sample and purified.

The specific activities *S* and S″ respectively are measured. For the first sample, the amount of unknown substance *m′* as given by equation (3) is

$$m' = \frac{mS}{S' - S} \text{ or } S' = \frac{mS}{m'} + S \quad ...(4)$$

and for the second sample,

$$m' = \frac{m''S'}{S' - S'} \text{ or } S' = \frac{m''S'}{m'} + S'' \quad ...(5)$$

By equating equations (4) and (5), we get

$$\frac{mS}{m'} + S = \frac{m''S'}{m'} + S''$$

or $$m' = \frac{mS = m''S''}{S'' - S} \quad ...(6)$$

The nature of the double isotope dilution method does not allow the degree of precision attainable with the direct and inverse isotope dilution methods. Thus, double isotope method is not an accurate method

APPLICATIONS OF ISOTOPE DILUTION METHOD

(i) A classical application of isotope dilution method is found in the work of Hevesy and Hobble. The gravimetric form used for the determination of PbO_2 was obtained by anodi c-oxidation after a long procedure involving many operations in which variable losses of lead may probably occurred.

Adding a known activity of the radioisotope lead-210 to the initial solution and measuring the radioactivity of the final weight of lead dioxide, Hevesy and Hobbie were able to determine the yield of the lead recovery and therefore, to calculate the total initial amount of the element, which in some of the determinations was more than two times the quantity recovered.

(ii) Zinc, copper, mercury, indium and other cations have been determined by this method as low as 10^{-9} to 10^{-10} gm in model experiments.

(iii) The isotope dilution principle has been applied to determine the volume of water in the body by using the radioactive tritium as a tracer.

(iv) A useful application of the isotopic dilution method has been made in connection with the extremely difficult and tedious process of analysing the mixture of amino acids resulting from the chemical hydrolysis of protein outside the body by using labelled amino acids containing N-15 to act as the tracer.

(v) The determination of γ-hexachlorocyclohexane by means of the Cl-36 labelled tracer of benzyl penicillin with the help of C-14 and of vitamin B_{12} with cobalt-60 are a few of the many examples where isotope dilution methods have contributed significantly in improving the analytical procedure.

STUDY OF REACTION MECHANISM

(a) Mechanism of Photosynthesis in plants : The process involves the production of sugar by the interaction of carbon dioxide and water in the presence of sun-light and chlorophyll. A small quantity of radioactive CO_2* containing radioactive oxygen (O^{17}) is mixed with ordinary carbondioxide and process carried out. It has been found that oxygen gas evolved along with sugar formation is non-radioactive.

This means that oxygen, liberated as gas, has come from water which contains non radioactive oxygen and not from carbon-dioxide. Therefore the reaction mechanism

$$6CO_2 + 6H_2O \longrightarrow C_6H_{12}2O_6 + 6O_2$$

is wrong. The correct mechanism should be

$$6CO_2{}^* + 12H_2O \longrightarrow C_6H_{12}O_6{}^* + 6H_2O|^* + 6O_2$$

(b) In Studying the Hydrolysis of Ester : By labelling oxygen, the mechanism of ester hydrolysis has been studied. The course of ester hydrolysis can be studied by using water labelled with O^{18}. The hydrolysis of an ester by water enriched with heavy oxygen is indicated as :

From the above it is found that it is the acid and not alcohol produced which is radioactive showing the mechanism of the reaction as shown above.

By using labelled carbon (C*), the nature of chemical bond can be detected. The exchange of some complex ions with oxalate ions labelled with C* proceeds in the following way :

(i) $[Co(C_2O_4)_3]^{3-}$ ... Slow or zero
$[Cr(C_2O_4)_3]^{3-}$ ... Slow or zero] Complexes were covalent

(ii) $[Fe(C_2O_4)_3]^{3-}$ ... Instantaneous
$[Al(C_2O_4)_3]^{3-}$ ... Instantaneous] Complexes were ionic.

(c) In studying the Chemical Linkages : One of the first application of tracer chemistry to structural problems was the establishment of the structure of thiosulphate ion.

When sulphur is heated in the presence of labelled sulphite ion, the thiosulphate ion is produced according to the following reaction :

$$S + \overset{*2}{SO_3} \rightleftharpoons S + \overset{*2-}{S_2O_3}$$

When this thiosulphate containing a labelled sulphur atom is broken down in an acid solution, the following reaction occurs

$$\overset{*\ 2-}{S_2O_3} \rightleftharpoons S + \overset{*2-}{SO_3}$$

This indicates that the bonding of the labelled sulphur is not affected either in the synthesis or in the decomposition of the thiosulphate. Thus, the structure of thiosulphate can be denoted as.

$$\left[\begin{array}{c} \quad\ \ O \\ \quad\ \ | \\ S=S=O \\ \quad\ \ | \\ \quad\ \ O \end{array}\right]^{2-}$$

In a similar type of experiment, it has been proved that the two lead atoms in lead sesquioxide Pb_2O_3 are not equivalent.

Radiocarbon- 14 has been successfully utilised as a tracer in studying mechanism involved in alkylation, polymerisation, catalytic cracking as well as many other reactions of industrial importance.

(d) Esterification : By using isotopic tracers it was possible to demonstrate that in the esterification reaction the C–OH bond in acid and H–O bond in alcohol undergoing cleavages give rise to ester with elimination of a water molecule as in

$$RCH_2COOH + HO^*CH_3 \longrightarrow RCH_2COO^*CH_3, + H_2O$$

The labeled atom (O^{18}) is shown with an asterisk.

(e) Mechanism of the Friedel-Craft's reaction : In the presence of anhydrous aluminium chloride as a catalyst, a large number of reactions known as the Friedel Crafts' reactions take place, for example,

$$C_6H_6 + ClCOCH_3 \xrightarrow{AlCl_3} C_5H_5COCH_3 + HCl$$

The mechanism of the reaction was established by using $AlCl_3$ labeled with Cl^{36} where the following steps involved leading to the products :

$$\left.\begin{array}{lll} CH_3COCl & \longrightarrow & CH_3CO^+Cl^- \\ C_6H_6 & \longrightarrow & C_6H_5^- + H^+ \end{array}\right\} \text{Dissociation}$$

$$AlCl_3^* + Cl^- \longrightarrow AlCl_4^* \quad \text{Catalyst–ion combination}$$

$$\left.\begin{array}{lll} C_6H_5^- + COCH_3^+ & \longrightarrow & C_6H_5COCH_3 \\ AlCl_4^* + H^+ & \longrightarrow & AlCl_3^* + HCl^* \end{array}\right\} \text{Ion–ion interaction}$$

This indicates that the product HCl carries one-fourth of-the total activity and the catalyst the rest. The ratio of activity of HCl to that of the residual catalyst should thus be 1/3. This is exactly what is noticed experimentally.

(f) Oxidation of fumaric acid by $KMnO_4$: A molecule of fumaric acid on oxidation by acidified $KMnO_4$ gives rise to carbon dioxide, water and formic acid as per the reaction,

$$\begin{array}{l} COOH \\ | \\ CH \\ \| \\ CH \\ | \\ COOH \end{array} + 5\,[O] \rightarrow 3\,CO_2 + H_2O + HCOOH$$

Now question arises, which of the four carbon atoms (two carboxylic and two methylene) in fumaric acid appear in CO_2 and which in formic acid ? This was investigated by using C^{11} in two carboxylic acids that the activity was completely associated with the CO_2 and none with the formic acid, indicating that the one of the methylinic carbons appears in the acid. The mechanism is

(g) Oxidation of CO by air in the presence of the catalyst MnO_2 : The oxidation of CO by air in the presence of MnO_2 takes place as per the reaction,

$$CO + MnO_2 \longrightarrow CO_2 + MnO$$

$$MnO + O\ (air) \longrightarrow MnO_2$$

By using MnO_2 labeled with O^{18} it was revealed that none of the enrichment in O^{18} remained in

CO_2 but wholly in MnO_2 indicating that oxidation of CO takes place directly in the presence of the catalyst, MnO_2 as in

$$CO + O\ (air) + [MnO_2^*] \longrightarrow CO_2 + [MnO_2^*]$$

(h) Decomposition of H2C>2 by PbO_2 : H_2O_2 decomposes in the presence of PbO_2 as per the reaction :

$$PbO_2 + H_2O_2 \longrightarrow PbO + H_2O + O_2$$

Now question is whether the molecular oxygen formed comes wholly from H_2O_2. By using O^{18}–labeled H_2O_2 it was revealed that the resulting molecular oxygen comes wholly from H_2O_2 as in the reaction :

$$O = Pb = O + H - O^* - O - H \longrightarrow PbO + H_2O + O_2^*$$

(i) Electrodeposition of Chromium : A solution of chromate Cr(VI) is used as the electrolyte in the electrolytic bath for the electroplating of metals by chromium. Since the electrodeposition of chromium involves the reduction of Cr(VI) to Cr(0), now question is posed whether Cr(VI) is directly reduced to Cr(0) or through Cr(III) as an intermediate.

The question was answered by using Cr^{51} in the electrolyte. For the electrodeposition purpose, a mixed electrolyte solution of CrO_4^{2-} and Cr^{3+} ions was used in which either CrO_4^{2-} ion was labeled with Cr^{51}. This is possible because isotopic exchange between CrO_4^{2-} and Cr^{3+} ions is absent at room temperature. It was shown that the resulting metal deposit contained the activity only when CrO_4^{2-} was labeled but not when Cr^{3+} was labeled suggesting that Cr(VI) is reduced directly to Cr(0) without the Cr(III) as an intermediate.

(j) Structure of PCl_5 and sodium thiosulphate : From the isotope exchange reactions, we learnt that the five chlorine atoms in phosphorous penta-chloride and the two sulphur atoms in thiosulphate ion are not equivalent. This lends support to the fact that when two or more atoms of the same element are present in a molecule, the structural equivalence of the atoms can be revealed by tracer technique.

Thus, the synthesis of PCl_5 from Cl_3 and Cl^{36}–labeled Cl_2 *i.e.,*

$$PCl_3 + Cl_2^* \longrightarrow PCl_5^*$$

and its hydrolysis giving rise to phosphorous oxytrichloride and hydrochloric acid as in

$$PCl_5^* + H_2O \longrightarrow POCl_3 + 2\ HCl^*$$

indicate that two Cl atoms in PCl_5 are different from the rest of the three Cl atoms as revealed from the activity measurements that remained wholly with the resulting HCl. This is in agreement with the trigonal bipyramidal structure accepted for PCl_5 with three Cl atoms in the equitorial plane and two along the vertices.

Similarly, the non-equivalent structure of the two sulphur atoms in thiosulphate ion can be established by means of tracer technique. The synthesis of sodium thiosulphate from a boiling solution of sodium sulphite with S^{36}–labeled sulphur, *i.e.,*

$$Na_2SO_3 + S^* \longrightarrow Na_2S_2^*O_3$$

and its decomposition to silver sulphide and other products by adding silver nitrate solution acidified with nitric acid, *i.e.,*

$$Na_2S_2^*O_3 + 2AgNO_3 + HNO_3 \longrightarrow Ag_2S^* + Na_2SO_4 + \text{products}$$

shows that the two sulphur atoms in thiosulphate ion are not equivalent as revealed from the measurement of the activity that remained wholly with the silver sulphide. Thus, the structure of the thiosulphate ion can be denoted as

(k) Mechanism of Tautomerism : The isotopic exchange is extensively used to study tautomerism. It is known that the exchange of hydrogen between organic compounds and water under ordinary conditions proceeds with the participation of only mobile hydrogen atoms entering into the composition of OH, NH, SH groups, but it does not involve the hydrogen atoms attached to the carbon. That there is no exchange of hydrogen between glyoxal and water proves the absence of tautomerism for glyoxal, which otherwise would lead to .the appearance of an OH group:

Here and in schemes that follow the crossed dashed lines signify the absence of reaction. The absence of tautomerism for acetoxime is shown in an analogous way :

In this case only one hydrogen atom is exchanged, and in the presence of tautomerism all the seven atoms of hydrogen must be involved in an exchange reaction. That there is no hydrogen exchange between the nucleus and the CH_3 group of toluene is a proof that the expected tautomerism is in fact absent : The absence of tautomerism for phosphorous acid has been proved by the fact that only two hydrogen atoms are involved in hydrogen exchange with water :

And in the case of hypophosphorus acid all the three atoms of hydrogen are exchanged owing to tautomerism : It has been shown that when a mixture of quinone and nucleus- deuterated hydroquinone is heated, there occurs a kind of transfer of heavy hydrogen from hydroquinone to quinone. This may be ascribed to the tautomerism of the two compounds. The process may be shown schematically as follows:

(I) Mechanism of Rearrangements : The mechanism of rearrangements has been extensively investigated by the tracer method. Use may also be made of isotopic exchange, a study into the intramolecular distribution of a tracer in the original compound and in the rearrangement product, and of the introduction of an additive analogous in structure to the substance under study, but containing tracer atoms. The rearrangement of l-naphthylamine-2-sulphonic acid into l-naphthyl-2-sulphonic acid in the presence of $Na_2^{35}SO_4$ in the solution does not result in the incorporation of ^{35}S into the molecules of naphthylaminosulphonic- acid.

This is evidence of the intramolecular mechanism of the rearrangement since otherwise ^{35}S would be incorporated by isotopic exchange into the organic molecule. The benzidine rearrangement has been investigated by introducing a similar reagent. To study the rearrangement to the initial substance there was added a similar substance labelled in the CH_3 group.

This process did not involve the formation of the crossed-rearrangement product: which could be formed by the intermolecular mechanism. Hence, the rearrangement occurs intramolecularly rather than by way of free radicals. The same method has been used to prove the intermolecular mechanism of hydrogen rearrangement on heating of phenols and their derivatives with addition of a similar compound :

The most popular method or examining the mechanism of molecular rearrangements is a study into the intramolecular distribution of tracer atoms. The pinacoline rearrangement proceeds by either of the two paths:

After the rearrangement product HIO is oxidized an inactive compound CH_3I is formed. Since HIO oxidizes the CH_3 group attached to the carbonyl group, it follows that the first scheme is correct. As an interesting example, a study into the rearrangement of naphthyl-1-sulphonic acid may be cited, as a result of which the sulphonic acid groups may occupy four different positions relative to the tracer carbon of the molecule, which proves the intermolecular mechanism of this process via the formation of a free radical, $\overset{*}{C}_{10}H_7$

A combined method has been used for investigating the Claisen rearrangement of allyl phenyl ether on heating. The ether is mixed with deuterated phenol, and the rearrangement product-allylphenol-is found to contain no deuterium, which suggests that the mechanism is intramolecular. Further, it has been found that the radioactive carbon introduced into the open hydrocarbon chain at the extreme position changes its place after the rearrangement to the ortho-position and remains in the initial position after the rearrangement to the para-position. This occurs due to the closure of the ring at the ortho-position in the former case and due to the double transfer, first to the meta-and then to the para-postion, in the latter case :

The intermolecular mechanism of rearrangement can also be established by studying the reaction rate as the substance is being diluted In this way, there has been found that hydrogen rearrangement of phenol on heating obeys a bimolecular law and depends on the second-power concentration of phenol.

(m) Mechanism of Isomerizotion : The principal method of investigating isomerization is the incorporation of a tracer into the substance responsible for isomerization. Butylene undergoes isomerization under the action of orthophosphoric acid. If orthophosphoric acid is labelled with tritium, then tritium is found to be incorporated upon isomerization into the methyl group. This process may be conceived as follows :

$$CH_2 = CH - CH - CH_3 \rightarrow CH_2 - CH - CH - CH_3$$

```
CH2 = CH – CH – CH3  →  CH2 – CH – CH – CH3
 |         |             |
 |         |            3H
3H — O ... H
HO – P –   OH           →  H3PO4
     ||
     O
```

Orthophosphoric acid directly participates in the isomerization reaction. In contrast to this, the isomerization of α-bromonaphthalene and β-bromonaphthalene in the presence of $ZnBr_2$ labelled with radioactive bromine proceeds without the formation of an intermediate complex since no transition of radiobromine from $ZnBr_2$ to β-bromonaphthalene is observed.

(n) Mechanism of Reactions Involving free-radical formation : A number of substances readily enter into reactions with free radicals. Such substances (radical acceptors) at rather low concentrations, of the order of 10^{-2} mole per litre, entrap radicals completely. In a liquid medium, halogen molecules are capable of functioning as acceptors of organic radicals. If halogen molecules contain a radioactive tracer, the products of their reactions with radicals will also be radioactive. This affords the possibility of following the formation of free radicals in any reaction, for example, in radiolysis and photolysis.

One of the polymerization mechanisms is a so-called free-radical or radical polymerization. Two paths of the formation of polymers by the free-radical mechanism can be vizualized.

1. $\overset{*}{R} + -CH_2 - \overset{|}{C} = \overset{|}{C} - \rightarrow - CH_2 - \underset{|}{\overset{|}{C}} - \underset{|}{\overset{|}{C}} - \overset{*}{R}$

2. $\overset{*}{R} + -CH_2 - \overset{|}{C} = \overset{|}{C} - \rightarrow - RH + -CH - \overset{|}{C} = \overset{|}{C}-$

$-CH_2 - \overset{|}{C} = \overset{|}{C} - + - CH_2 - \overset{|}{C} = \overset{|}{C}- \rightarrow \overset{|}{C} = \overset{|}{C}-\overset{|}{C}H-\underset{|}{\overset{|}{C}}-\underset{|}{\overset{|}{C}}-CH_2 -$

In the first case each unit in the polymer contains an initiating radical, and in the second, these radicals form end groups, in a small amount, in the polymer chain. If the initiating radical has a radioactive tracer, the specific activity of the initial radioactive compound is compared with that of the polymerization product. Such a comparison demonstrates that the polymerization proceeds by the second mechanism.

(o) Mechanism of catalysis : The schemes of catalytic reactions presuppose, in a number of cases, the direct participation of a catalyst in the reaction as a carrier of atoms. Such a catalyst is, for example, aluminium bromide, in whose presence there takes place a rapid exchange of bromine between two alkyl bromides. This can easily be verified by introducing alkyl bromide labelled with radioactive bromine into the reaction and observing the change of the activity of the components of the mixture with time.

$$R'\overset{*}{Br} + Al_2Br_6 \rightleftharpoons R'Br + Al_2Br_5\overset{*}{Br}$$

$$R''Br + Al_2Br_5\overset{*}{Br} \rightleftharpoons R''\overset{*}{Br} + Al_2Br_6$$

As an example of the investigation of a catalytic process by the radioactive tracer technique, we may cite the processing of natural methane into a mixture of hydrocarbons. It is conducted by successive conversion of methane into water gas :

$$CH_4 + H_2O \longrightarrow CO + 3H_2$$

with subsequent synthesis, from the water gas, of hydrocarbons over a nickel or iron catalyst. It was believed that in this process a metal carbide is first formed on the catalyst, and then the carbide with hydrogen forms a mixture of hydrocarbons :

$$2Ni + CO \longrightarrow NiO + NiC$$

$$NiO + H_2 \longrightarrow Ni + H_2O$$

$$NiC + H_2 \longrightarrow Ni + CH_2$$

$$n\overset{*}{C}H_2 \longrightarrow C_nH_{2n}$$

$$C_nH_{2n} + H_2 \longrightarrow C_nH_{2n+2}$$

A study of this reaction was performed with the aid of radioactive carbon incorporated, in the form of a carbide, into the catalyst. In this case, it was found that the hydrocarbons do not practically contain radioactive carbon and the catalyst is not a direct carrier of carbon atoms; so the scheme given above proves to be incorrect.

It has been shown, in an analogous way, that the oxidation of carbon monoxide to carbon dioxide over manganese dioxide labelled with heavy oxygen proceeds without the participation of the heavy oxygen, and carbon dioxide obtained has the natural isotopic composition of oxygen:

$$CO \xrightarrow[MnO_2^*]{O} CO_2$$

An investigation of the stages involved in a catalytic reaction can be performed by means of the methods described above (through a study of the intramolecular distribution of the activity; isotopic exchange; Neiman kinetic method).

An example of a study of the mechanism of catalysis by the Neiman kinetic method is the oxidation of ethylene over silver. This process may proceed through the formation of ethylene oxide or acetaldehyde. If a mixture of ethylene labelled with carbon-14 and ethylene oxide is subjected to oxidation, the observation of the change in the radioactivities of ethylene, ethylene oxide and carbon dioxide will show that the specific activity of ethylene is equal to that of the carbon dioxide and is much higher than the specific activity of ethylene oxide, which indicates that the last named compound is not an intermediate reaction product. Experiments with addition to the mixture of acetaldehyde have demonstrated that its specific activity undergoes change with time, so that we can conclude that it is an intermediate product of oxidation.

A study into the mechanism of chemical reactions as a function of the small concentrations of impurities introduced into the catalyst is of great value. For example, with the aid of ^{65}Zn it has been established that if the catalyst (ZnO) contains as low as 0.18 per cent of $ZnSO_4$, the decomposition of isopropyl alcohol proceeds to the side of formation of propylene and water, whereas in the absence of $ZnSO_4$ the predominant reaction is the formation of acetone and hydrogen.

The change in the concentration of phosphate in a palladium catalyst drastically alters its catalytic properties. This has been illustrated by the decomposition of hydrogen peroxide in the presence of phosphorus-32 in the impurity. The rate of decomposition sharply increases up to concentrations of phosphorous of the order of 10^{-3} per cent, and then falls off abruptly.

The isotopic exchange of oxygen between an oxide catalyst and molecular oxygen serves as a characteristic of the activity of the catalyst since it changes in the same direction as the mobility of oxygen in the catalyst. When promoters (K_2SO_4) are introduced into vanadium oxide, the rate of oxygen exchange is increased.

Radiochemical methods enable one to determine the surface area of a catalyst and the degree of its surface homogeneity. The homogeneity of the surface of catalysts is determined by means of the differential isotopic method worked out by S. Z. Roginsky. In this method, first a small amount of a labelled substance is adsorbed on the surface, and then the same but unlabelled substance is adsorbed until saturation is reached. After this, slow desorption is carried out and the specific activity of the desorbed substance is measured. If the surface is homogeneous, the specific activity remains constant during the desorption, and in the case of a nonhomogeneous surface the specific activity gradually increases.

(p) Mechanism of corrosion : A study of the wet corrosion of iron by the tracer method has shown that at first the oxidation of iron proceeds at the expense of labelled water oxygen. In the presence of salts the effect of ions increases in the following sequence :

$$ClO_4^- < F^- < SO_4^{2-} < I^- < Br^- < Cl^-$$

As has been found with the aid of copper-64 the oxidation of copper proceeds by way of diffusion of copper through the copper oxide film and the oxidation of diffusing atoms.

10. Age of the Earth : Here we make use of the fact that uranium and thorium decay to give the end products which are isotopes of lead and are stable. With the assumption that the radioactive minerals were formed along with the rocks containing them, it is possible to find the number of atoms decayed which is related to the time the minerals took to decay to their present state of stable isotopes and is a measure of the age to the rocks.

In the usual lead-uranium method, rock containing the nuclides ${}_{92}U^{238}$, ${}_{92}U^{235}$, ${}_{82}Pb^{206}$ and ${}_{82}Pb^{207}$ is chosen. The isotope of ${}_{82}Pb^{208}$ which follows from the decay of thorium is omitted by selecting such a rock which does not contain thorium. Of these ${}_{82}Pb^{208}$ is non-radiogenic since it has no place in any of the four radioactive series. The isotopes

Pb^{206} and Pb^{207} follow from U^{238} and U^{235} nuclides respectively. For the $U^{238} \rightarrow Pb^{206}$ decay we can write :

$$N_{206} = N_{238} (e^{\lambda_1 t} - 1) \qquad ...(1)$$

where N_{206} is the nuclides of stable isotope of ${}_{82}Pb^{206}$, N^{238} is the present-day number of nuclides of ${}_{92}U^{238}$, λ_1 is the decay constant of U^{238} while *t* is the age of the rock.

Similarly for $U^{235} \frac{N_{204}}{N_{207}}$ Pb^{207} decay we have

$$N_{207} = N_{235} (e^{\lambda_1 t} - 1) \qquad ...(2)$$

From equations (1) and (2), we have

$$\frac{N_{206}}{N_{207}} = \frac{\overset{*}{N_{238}}}{N_{235}} \left[\frac{e^{\lambda_1 t} - 1}{e^{\lambda_2 t} - 1} \right] \qquad ...(3)$$

The present-day abundances of N_{238} and N_{235} are 137.8 : 1 as measured on the mass spectrometer.

$$\frac{N_{206}}{N_{207}} = 137.8 \left[\frac{e^{\lambda_1 t} - 1}{e^{\lambda_2 t} - 1} \right] \qquad ...(4)$$

Also $T_{1/2}$ $U^{238} = 4.5 \times 10^9$ years giving $\lambda_1 = 4.87 \times 10^{-18}$ per sec and $T_{1/2}$ $U^{235} = 7.13 \times 10^{18}$ years giving $\lambda_2 = 3.88 \times 10^{-17}$ per sec.

If the present day lead ratio $\frac{N_{206}}{N_{207}}$ is determined on the mass spectrometer, then *t* can be found out.

The method is valid for rocks containing pockets of uranium lead mineral.

The *age of earth* is measured by measuring the relative abundances of lead isotopes in lead ores such as galena which have ceased to be radioactive. When the ore was first formed t_0 year ago (from now) in the earth's crust where t_0 is the age of the earth, the isotopic composition of all kinds of lead was the same and the Pb^{204} isotope contents have remained constant ever since as this isotopes non-radiogenic. If the total Pb^{206} (from the decay of U^{238}) content as measured today on a lead sample is α times the Pb^{204} contents, then this includes the original non-radiogenic portion *a* and the radiogenic portion N^{206} given as

$$N_{206} = \alpha - a, \frac{N_{206}}{N_{204}} - \alpha$$

The $\frac{N_{206}}{N_{204}}$ is measured accurately by the mass spectrometer.

Similarly if the total of Pb^{207} (from the decay of U^{235}) is measured to be β times the Pb^{204}, then also, if *b* is the non-radiogenic portion in it, we have

$$N207 = b - b$$

$$\text{Hence} \frac{\alpha - a}{\beta - b} = \frac{N_{204}}{N_{207}} = 137.8 \frac{[e^{\lambda_1 t} - 1]}{[e^{\lambda_2 t} - 1]} \quad ...(5)$$

In eq. (5) t= decay time 'of rock

= total age of the rock (age of the earth)

– (age since rock fully mineralized into lead isotope),

all times measured from the present time.

$= t_0 - t$

Eq. (5) can thus be written as

$$R = \frac{\alpha - a}{\beta - b} = 137.8 \left[\frac{e^{\lambda_1 t_0} - e^{\lambda_1 t}}{e^{\lambda_2 t_0} - e^{\lambda_2 t}}\right] e^{(\lambda_2 - \lambda_1) t} \quad ...(6)$$

From the above equation, it follows that uranium started decaying to years ago and was completely transformed into radiogenic lead *t* years ago as measured from now. Since then the ratios of different isotopes have remained constant and quantity of remaining, if any, is negligible.

The method of calculation consists in picking galenas from rocks with same mineral age *t* and assuming that all non-radiogonic leads have same isotopic composition so that, for instance for, two galenas

$$a_1 = a_2 = a \text{ and } b_1 = b_2 = b$$

Following eq. (6), we have

$$\alpha_1 - a = k(\beta_1 - b)$$

etc. $\quad \alpha_1 - a = k(\beta_2 - b)$

Taking *a* and *b* as constants we can plot values of a and P for several galenas of the same mineral age *t* and the graph will be linear of having the slope.

$$k = 137.8 \left[\frac{e^{\lambda_1 t_0} - e^{\lambda_1 t}}{e^{\lambda_2 t_0} - e^{\lambda_2 t}}\right] e^{(\lambda_2 - \lambda_2) t}$$

The above result will give the value of *to* if *t* is known which is common to all the rocks measured. The most probable age of the earth by this method comes out to be $(4.55 \pm .07) \times 10^9$ years.

11. Radiometric Titrations: During the course of radiometric titrations, radioactivity is the *property* of the solution that is followed for the location of the end point. Under similar conditions of measurement, the amount of radioactivity is found to be proportional to the concentration of the reacting species which results in identical changes in the radioactivity.

The end- point of the reaction is obtained by plotting the changes in the radioactivity against the amounts of titrant added. Because a linear relation exists between concentration of the labelled substance and radioactivity, the titration curve is also linear. The end-point is obtained by the point of intersection of the two linear portions of the titration curve.

Radioactive P^{32} was converted to a soluble phosphate and added to a standard solution of disodium hydrogen phosphate. This solution was used to titrate several substances such as :

Ba (II), Pb (II), Th (IV) and Mg (II).

After each addition of phosphate solution, a sample of the clear filtered solution was suck up around a G.M. tube and the activity was measured. The activity remained essentially constant until the equivalence point was reached. At the end point it rose rapidly with addition of the reagent. From the intersection of the activity curves the end point was accurately determined.

Direct titration : If a substance produces a radioactive precipitate upon titration with a radioactive titrant, the end-point of the reaction will be indicated by :

(i) the inflection point of the activity curve for the precipitation, the activity of which will be plateau at the end point,

(ii) the inflection point of the activity curve for the supernatant, which will show an abrupt increase at the end point.

In this type of titration, measurements of activity are made after each addition of titrant, and the end point is determined through the construction of the curve

Indirect titration : In this type of titration, a measured excess of titrant is added to the sample. After mixing the solution it is centrifuged and the activity of an aliquot of the supernatant is determined. A comparison of activity of the supernatant to that of the original titrant measures the concentration of unknown solution.

Advantages

In radiometric titrations, the end-point can be established much more accurately than with the conventional coloured organic indicators. The method of radiometric titration readily lends themselves for use in automatic analysis. Furthermore, the titrations can be carried out in heterogeneous, coloured, turbid, corrosive or non-aqueous system because radioactivity is not affected by these conditions.

(a) The quantities required for estimation are in the order of 10–15 mg but can be applied to still lower amount *e.g.* in few μg amount.

(b) Weighing is not required and chemical purity is not considered.

Disadvantages of Radioactive Titrations

(a) Radioactive chemical purity of the reagent and the precipitates formed is essential.

(b) The error in the method can never be less than the counting error, which is ordinarily 0.5%.

(c) It is applicable in the cases where the stability of precipitate is low and extraction coefficient is high.

The principle and techniques of the radiometric titrations in details have been reviewed by **Bruan** and **Tolgyessy** in 1914.

TYPES OF RADIOMETRIC TITRATIONS

A. Based upon precipitate formation : In such titration the radioactive indicator is precipitated during the titration The end-point can be located by following the changes in the radioactivity of one of the phases. In these titrations, the appropriate isotopes of the element or the elements forming the compound to be determined are employed.

Radionuclides applied as indicators should meet the following requirements :

(a) Half-life : The use of radionuclides with a half-life ranging from a week to several months has proved to be most suitable.

(b) Nature and energy of the emitted radiation : The radionuclides emitting gamma rays are commonly used in radiometric titrations.

(c) Chemical form of indicator : The chemical form of radioactive indicator should be mostly the same as that of the substance to be labelled.

The radiometric titrations based on precipitate formation may yield three types of titration curves. The shapes of these curves will depend on whether the titrant, the test solution or both are labelled. In few cases when no appropriate radioisotopes of the element to be determined are available then radioactive isotope of another element can often be used for the indication of the element to be determined. The technique is denoted as the *non-isotopic method* of labelling.

B. Based on Complex Formation : In the radiometric location of the end-point of complex-formation reactions, the separation of the components is the most difficult operation. In such reactions, the reacting species and the reaction products are in the same phase, before and after the reaction. Therefore, such titrations can only be carried out by employing an auxiliary method of separation. Several separation methods such as solvent extractions paper chromatography, ion exchange etc, have been reported. The sensitivity of such titrations depends on factors known to operate in complexometry. However, in the present case sensitivity is affected by the method of separation as well.

C. Based on redox reaction : As in complexometric radiometric titrations, the difficulty in the location of the end-point of radiometric redox titrations is the separation of component. This can be solved only by introducing an auxillary separation process. The use of solid indicators has been reported.

Radiometric titrations in non-aqueous media : Non-aqueous solvents are used mainly in the titrimetric determination of acids and bases. Reactions based on precipitate formation, redox reactions and other reactions have been utilised in non-aqueous titrations. This stems mainly from the difficulties encountered in applying the electrochemical and other methods of end point detection in non-aqueous solvents. Accordingly, *the potential introduction of radiometric end-point detection* in this field appears to be promising.

The theoretical and practical fundamentals of radiometric titrations based on precipitate formation carried out in non-aqueous media have been given by Chernyl and others.

The sensitivity limits of titration based on precipitate formation are known to be determined by the solubility product of the precipitate formed during titration.

Applications of radiometric Titrations : Radiometric titrations have been applied with success in fields other than direct analysis.

(i) ***Determination of the composition of compound*** **:** Radiometric titrations have been applied successfully in the determination of the composition of compounds formed during titration in several instances. For example, the composition of the triphenyl selenium tetraiodobis-muthate (III) complexes have been elucidated by Shinagawa and others. However, radiometric titrations have been little used in this field.

(ii) ***Investigation of co-precipitation*** **:** The study of co-precipitation is of extreme importance chemically, particularly as regards nuclear chemistry, where co-precipitation has an essential role. Alimarin and Sirotina studied the co- precipitation of silver, thallium and lead with iodide, thiocyanate, sulphide using radioactive isotopes ^{240}Tl, ^{210}Ag and ^{212}Pb.

(iii) ***Determination of the specific activity of radioactive preparations***. The use of radiometric titrations in the determination of the specific activity of radioactive preparations has recently been suggested by Kametani and others. This possibility is favourable because the labelling of the test solution becomes dispensable. The radioactivity of the substance to be determined serves as its own indicator.

Carbon Dating

Carbon dating is the method to determine the age of the relics of distant past. This technique is used by historians, archeologists, anthropologists, geologists, palaeontologists and palaeobotanists. It was **Willard F. Libby** (1954) who developed this technique.

Theory : Carbon has an atomic weight of 12. Radioisotope of carbon with an atomic weight 14 is significant for carbon dating. Radiocarbon (C^{14}) is produced in the upper atmosphere by the transmutation of nitrogen atom under the influence of cosmic rays (free neutron).

$$_7N^{14} + {}_0n^1 \longrightarrow {}_6C^{14} + {}_1H^1$$

$$_7N^{14} + 3\ {}_0n^1 \longrightarrow {}_6C^{14} + {}_1H^3$$

The total amount of radioactive carbon-14 in our earth remains constant. The disintegration of radioactive carbon-14 forms nitrogen back from it :

$$_6C^{14} \longrightarrow \beta + {}_7N^{14}$$

Carbon-14 may enter into the formation of atmospheric carbon dioxide gas. CO_2 is absorbed by plants during photo-synthesis and is later incorporated into their bodies. Animals too consume C^{14} by eating plants. On death organisms cease to take in fresh carbon atoms. Carbon-14 begins to decay. Half-life of C^{14} is 5568 years. After 5568 years a fossil (plant or animal) will lose half the amount of carbon–14 present in its living state. The amount of C^{14} in any ancient organic sample may, thus, tell its age.

Methodology : Method of carbon dating is simple but requires a well equipped laboratory and sophisticated equipment.

(i) The sample wood, bones or other types of organic remains are first cut into smaller chips.

(ii) The material is placed in heating tube for converting it first into carbon and then into CO_2.

(iii) The gas is purified and finally frozen to solid and stored.

(iv) Geiger counter is employed to determine the rate of emitting fundamental particles from the frozen CO_2.

By suitable calculation, the age of sample tested can be worked out.

Limitations : Age determination by carbon dating is prone to errors. Errors develop because of the variability in the cosmic rays output from the sun. Variation in the availability of carbon-14 at a particular place in different times and the fact that plants have a special aptitude to incorporate varying amounts of carbon-14 in them.

By proper manipulation and experiments, these errors can be reduced to minimum.

Importance

1. Carbon dating has proved to be a great tool for correlating facts of historical importance.
2. It is very useful in understanding the evolution of life, and rise and fall of civilizations.

TECHNIQUES IN BIOLOGICAL SYSTEMS

Some of the most important isotopic tracer studies have been made with living organisms. One type of investigation is to determine the manner in which a particular ingested element distributes itself in various parts of the organism. For example, an animal may be fed an amino acid with labeled nitrogen, and then the amounts present in several organs, such as the kideys, liver? heart, and lungs and in the blood, may be determined after the lapse of a certain time.

This work can be carried a stage further, by a more detailed analysis which will reveal how much of the labeled nitrogen is present in the form of a particular amino acid in various parts of the organism. A search for intermediate compounds containing the isotopic tracer may throw light on the mechanism of the processes taking place. Similar observations can be made with other elements, such as carbon and oxygen, and with green plants and bacteria as well as with animals.

ISOTOPIC TRACERS IN CELLULAR BIOLOGY

(a) All living organisms are made up of cells and, as far as is known, all complete cells contain the two essential substances deoxyribonucleic acid (DNA) and ribonucleic acid (RNA). Almost all the DNA is found in the nucleus of the cell and, in fact, only in the chromosomes, the threadlike structures present in the nucleus. There is some RNA in the cell nucleus, but most is in the cytoplasm, the cell liquid which surrounds the nucleus. The DNA carries the genetic code (or instructions), which causes each generation to resemble the preceding one, and also information for the production of various proteins from the amino acids in the cell fluid. The RNA, of which three types with different functions are known, transfers the information from the DNA to the machinery in the cell that synthesizes proteins.

(b) In simple terms, the DNA molecule consists of two twisted chains made up of a sequence of units of the sugar-like compound deoxyribose attached to a phosphoric acid residue. The two chains are connected together at each of these units by a pair of organic bases, like the rungs of a ladder. Four bases are available to form these rungs, but it is found that only two particular pairs, of the six theoretically possible, are actually used. The structure of RNA is similar to that of DNA, with two exceptions : first, the sugar-like compound in the chain is ribose which contains an –OH group in place of a –H atom in deoxyribose; and second, three of the bases that form the rungs between the chains are the same as in DNA but the fourth is different. Actually, DNA and RNA are not specific compounds, but rather structural types that can exhibit a large number of variations depending on the order in which the pairs of bases are attached to the sugar- phosphate chains.

(c) The growth of an organism and the replacement of spent tissues is achieved as the result of individual cells each dividing into two identical cells; the process is called mitosis. Before this division can occur the amount of DNA in the original nucleus must be doubled, so that each new cell has the same quantity as its parent did when it was formed. A cell in which DNA is being generated is consequently one which will soon divide. It is possible to label the DNA by means of a radioactive tracer and thereby to study the behaviour of cells during mitosis.

(d) In the labeling of DNA use is made of its specific ability, not possessed by RNA, to take up the base thymine; the latter must be supplied in the form of a precursor called thymidine, which is a compound of thymine and the sugar deoxyribose. One of the hydrogen atoms in the thymine can be replaced by its radioactive isotope tritium, leading to the formation of tritiated thymidine (or ^{3}-thymidine). If this labeled thymidine is made available, *e.g.*, by injection into the bloodstream or by addition to the fluid medium containing the cells, it is incorporated into the DNA molecules being produced in the chromosomes of the cells that are preparing to divide. The location of the tritium, and hence of the newly formed DNA, in the cell nucleus is then detected by radioautography.

(e) The beta particles from tritium-there are no gamma rays-have very low energies (maximum 0.018 MeV) and consequently short ranges. It is necessary, therefore, to have close contact between the cells under examination and the sensitized photographic emulsion. This is done by placing some of the cells on a glass slide and dipping the slide into a liquid emulsion made of gelatin and small grains of silver bromide. Upon removal, a film of photographic emulsion, which soon sets, is formed on the slide. This is kept in the dark in a refrigerator for the required exposure time, usually several days. The film is then developed and fixed in the normal manner. The physical outlines of the cell and its nucleus are rendered visible by staining techniques used in conventional cell studies. The black dots of silver in the radio-autograph indicate the location of the tritiated thymidine in the DNA.

(f) As is well known to biologists, not all cells are able to undergo mitosis; with the tracing technique described above, it is possible to distinguish those cells that do divide from those that do not. It has been found in this manner, for example, that only about 3 percent of the cells in the adult human body are capable of dividing for the purpose of tissue repair. Furthermore, tritium tracer experiments have shown that in the cycle of a dividing cell, *i.e.*, the time elapsing between its initial formation and subsequent mitosis, DNA is produced during about half the time only.

The formation occurs in the latter part of the cycle, just before the actual division of the cell into two is seen to take place. The length of the cycle varies with the type of cell and the organism of which it is a part. It can range from about 8 hours to several days. (g) Malignant (cancerous) cells are characterized by the abnormally high rates at which they increase in number. Tracer experiments with tritiated thymidine have indicated that this is not due to the cells dividing more rapidly; in fact, the cycle time is frequently greater than for normal cells.

The increase in growth rate of the malignant cells is apparently due to the much larger proportion of the cells that are capable of further mitosis. In normal tissue, about half the cells formed are able to divide; this keeps the total number of cells almost constant. But in some cancerous tissue, nearly every cell can divide further and so the cell population increases more and more rapidly with time.

(h) A different type of application of tritium-labeled DNA is in the study of the cells produced in the blood-forming tissues, *i.e.*, the, bone marrow, lymph glands, and spleen. These tissues take up tritiated thymidine that has been injected in the organism, and radioactive DNA makes its appearance in the red and white blood cells; such cells are not able to divide and have a limited lifetime. By observing the changes with time in the amounts of radioactivity present in the different blood cells, data are obtained regarding the rate of production of the cells, their speed of migration into the bloodstream from the tissue in which they are formed, and their life spans. One of the unexpected discoveries is that the white cells called *lymphocytes*, formerly thought to have a short life, appear to have an average lifetime of at least 100 days. It has consequently become necessary to reconsider the physiological significance of the lymphocytes in the blood.

(i) There is, unfortunately, no specific precursor for RNA, such as thymidine is for DNA. Nevertheless, there are a few nucleosides, *i.e.*, amino acid-sugar compounds, which are incorporated preferentially, although not exclusively, in the RNA. With these tritiated compounds it has been established, using radioautography., that RNA is synthesized in the nucleus of the cell and is then slowly transferred to the surrounding cytoplasm.

The production of RNA by cells differs in important aspects from that of DNA; the former is produced almost constantly by the nuclei of nearly all cells, whether they are dividing or not. As a result, all such cells are capable of synthesizing proteins from amino acids.

(j) To obtain more detailed information about the characteristics of the three types of RNA that play a role in protein synthesis, a different procedure must be adopted. The nucleoside precursor is labeled with carbon-14, which emits more energetic beta particles than does tritium, so as to permit scintillation counting. The labeled RNA formed in the cells is'separated from the other cell constituents by chemical- methods, and is then split up, by centrifugation, into the three types by taking advantage of the differences in their molecular weights.

The formation of RNA (or of each of the three types) is followed by measuring the radioactivity of the carbon-14 present in the following manner. A known quantity of the RNA is added to a liquid scintillator and the rate of production of scintillations is observed with a photomultiplier tube. This rate is proportional to the amount of carbon-14 in the RNA. From such observations made at different times, the rate of production of the RNA can be determined.

(k) An outstanding contribution to the clarification of certain life processes has resulted from the work done by R. Schoenheimer and D. Rittenberg and their associates. It had long been accepted that degradative changes in the living animal took place slowly, the purpose of food being largely to supply the currently required energy, while a small proportion went to replace worn out tissue. As a result of experiments carried out with the aid of deuterium and stable nitrogen-15 as tracers, this concept of an essentially static or dormant state of the organism has been shown to be entirely wrong. The present view is that there exists, as Schoenheimer has called it, "a dynamic state of the body constituents," in which there is a continual interchange between the fats, proteins, and carbohydrates already present in the animal body and those ingested is the form of food.

(1) Linseed oil, which contains fats derived from doubly and triply unsaturated acids, was partially "hydrogenated" by means of deuterium, so as to yield a mixture of both saturated and slightly unsaturated fats, in which two or four of the hydrogen atoms attached to carbon were replaced by deuterium. The resulting deutero-fats, labeled in this manner, were fed to animals, and the surprising fact was observed that only a small proportion of the deuterium was excreted for several days. The major part of the deuterium was found to have been deposited in the fatty portions of the body. Even when the diet was very deficient in fat, and the total calorie supply was inadequate, so that the amimal was drawing upon its reseves, the deutero-fat was mainly stored and not put to immediate use. Obviously the view that the organism was in a more or less static state, using such food as was needed and storing the remainder, if any, was unsoud.

(m) After the natural diet of the animals had been resumed, the labeled fats were found to disappear gradually, the deuterium leaving the body in the form of water. But if the water included in the normal diet of other animals was enriched in heavy water, so as to maintain a constant level of deuterium in the body fluids, the stored fat was found to gain deuterium at the same rate as it was lost by the animals which had previously been fed the deutero-fats. These results indicated that reversible (dynamic) equilibria, involving fats and water, occur in the living organism. The saturated fats are "desaturated" by the removal of hydrogen (or deuterium) and at the same time the unsaturated fats are "saturated" by the addition of hydrogen (or deuterium) from water.

(n) Further investigations showed that it was only the singly unsaturated fats which could take part in this equilibrium. The more highly unsaturated fats, such as those derived from linoleic and linolenic acids, could not be saturated, neither could they be formed by removal of hydrogen from saturated fats in the body. This conclusion.reached unequivocally by labeling the various fats with deuterium and tracing their behavior in the animal, confirmed the opinion held by nutritionists that the highly usaturated fats are "indispensable," in the sense that they must be supplied in the diet, the body being unable to synthesize them.

(o) In the continuation of their studies on animal metabolism, Schoenheimer and Rittenberg prepared a number of amino acids in which the nitrogen atom of the amino ($-NH_2$) group was labeled with the nitrogen-15 isotope. Among the many interesting observations which were made after adding a labeled amino acid to the diet of an animal was the following : nearly all the amino acids isolated from the tissue protein contained nitrogen-15, but the concentration was greatest in the amino acid corresponding to the one which had been included in the diet. It would thus appear, first, that the dietary amino acid is taken up directly and rapidly into the body protein and, second, that there is a bilogical transfer of nitrogen from one protein amino acid to another during metabolism.

(p) As a consequence of the nitrogen-transfer reactions, the nitrogen fed to an animal in the form of one amino acid should eventually be distributed among all those present in the body protein. There is, however, one exception to the rule. The indispensable amino acid lysine does not seem to be able to take part in this exchange of the amino groups. Incidentally, lysine is also known to be unique in another respect; unlike the other indispensable acids, it cannot be replaced in the diet by the corresponding alpha-hydroxy acid. These two properties of lysine are undoubtedly related.

(q) By labeling various dietary amino acids with both isotopic nitrogen and deuterium it has been shown, further, that the formation of creatine in the body requires contributions from three amino acids, namely, glycine, arginine, and methionine. The creatine is produced in this manner, and is converted into creatinine, at a fairly steady rate. This acounts for the observation, which was misinterpreted, as indicated above, that there is an approximately constant excretion of creatinine independent of the amount of the dietary protein.

TRACER STUDIES OF BLOOD

(a) Some of the most valuable information in the biological field has come form investigations made in the United States with the isotopes of iron. As just indicated, this element differs from other elements in the respect that iron supplied in the food, or injected into the animal, does not exchange to any appreciable extent with iron already present in the body. From observations on the feeding of iron labeled with radioactive iron-59 to normal animals it appeared that relatively little of the iron found its way into the blood, where most of the iron in the body occurs as hemoglobin, the coloring matter of the red corpuscles (erythrocytes). In animals depleted of iron, however, a larger proportion of this element is absorbed. Shortly after the loss of blood, iron is not absorbed even though there is a deficiency of hemo- globin; but, in due course, as the hemoglobin is replaced by drawing upon the animal's store of iron, there is considerable absorption of iron from the food. Similarly, it has been shown that in cases of pernicious anemia iron is absorbed only after the reserves have been depleted by the administration of liver extract.

(b) The extent to which iron is taken up evidently depends on the state of the reserves of the element in the body. The view now generally held, based largely on studies with radioactive iron as a tracer, is that iron is stored in the body in the form of an iron-protein combination known as ferritin. When the store of ferritin has attained its normal value, it does not increase further, no matter how much iron may be supplied orally. If, for some reason, however, the iron reserves are drawn upon, so that the amount of ferritin is decreased, the body is able once more to absorb iron.

(c) The remarkable fact about iron, which makes its behavior so different from that of the other body elements, is that it is utilized over and over again. By employing radioiron as tracer, it has been found that iron remains in the hemoglobin as long as the red blood-corpuscles are intact. But when the corpuscles are destroyed the iron is not lost; it is almost wholly retained in the body, and is rapidly incorporated into the hemoglobin of new red corpuscles. It is this phenomenon which accounts for the small absorption of iron from the food under normal conditions, and the absence of exchange between administered and stored forms of the element.

(d) The stability of the iron in the red corpuscles has led to the development of a method for determining the total amount of red cells in the blood. A small quantity of iron, containing some iron-59, is injected into an animal under such conditions that the element is absorbed and forms hemoglobin. Some of the red corpuscles, labeled in the hemoglobin, are withdrawn and injected into the subject. As there is no appreciable exchange of iron, all the iron-59 in the injected blood remains in the hemoglobin. By comparing the radioactivity of the red cells in the subject's blood, after time has been allowed for thorough circulation, with that of the injected cells the extent of dilution can be calculated. If the quantity of red cells injected is known, the total amount in the subject can be determined by proportion. A simpler procedure than that just described is to take a sample of the patient's blood and label the red cells with chromium-51 (half-life 27.8 days) by adding sodium chromate outside the body. The labeled red cells are then separated and a known amount injected into the patient for the dilution study. The same general (dilution) principle is used extensively to determine the total blood volume of the patient, by injecting a known volume of blood labeled with chromium-51.

(e) One way of determining the average life of the red corpuscles of the blood is to make use of the stable nitrogen-15 isotope. Hemoglobin, which plays a vital part in the transport of oxygen in the body, is itself a combination of two units : one is the red pigment called hemin, and the other is a protein, known as globin. In hemin the element iron is present in association with a nitrogen-containing mplecule, belonging to the group of complex compounds referred to as porphyrins. From experiments made with animals which had been fed the simple amino acid glycine labeled with nitrogen-15, it was known that this substance is an important source of the nitrogen in the porphyrin, and hence in the hemin of the red corpuscles. In a continuation of this study, labeled glycine was fed to a human adult for three days; samples of blood were withdrawn, from time to time, during the ensuing months. The crystaline hemin was extracted from the blood, and its nitrogen-15 content determined by means of a mass spectrometer.

(f) It was observed that the amount of nitrogen-15 in the hemin increased steadily for about 25 days after feeding of the labled glycine had ceased; it then remained fairly constant for the succeeding 80 days or so, after which it declined steadily. The interpretation of these results is as follows. The glycine, after ingestion, is rapidly stored in the body, as is known to be the case from other experiments. During the course of some 25 days it is gradually released to form hemin which, in combination with globin, produces the hemoglobin of the red blood-corpuscles. Since the nitrogen-15 content then remains the same for some time, it is evident that the hemin contained in these corpuscles is not continuously, broken down and rebuilt. Once it has formed it has a fairly definite life span, estimated, from the change in the amount of isotopic nitrogen, to be about 120 days. At this age, a red corpuscle apparently disintegrates, releasing the hemin, which then breaks up into iron and the nitrogen-containing porphyrin. As seen above, the iron is reutilized, combining to form more hemin, but the porphyrin part passes out of the body, as is shown by the steady decline in the proportion of nitrogen-15 in the blood.

(g) A notable practical application ofradioisotopes in connection with the study of the characteristics of blood for transfusion makes use of both iron-55 and iron-59. The blood volume of the recipient, before transfusion, is first determined by introducing a small quantity of red cells labeled with iron-59, as explained above. Transfusions of whole blood including red cells labeled with iron-55 are then given. Since the two radioactive isotopes have different characteristics, they can be determined separately in the blood of the recipient. In this way, the fate of the transferred blood can be followed through the behaviour of the iron-55, whereas the concentration of iron-59 gives the effective blood volume. As a result of the knowledge obtained in this manner, valuable advances have been made in the storage of blood.

MEDICAL APPLICATIONS

The innumerable uses of radioactive tracers in medicine and biochemistry are known. Within the limits of the present book, it will not be possible to give even a superficial review of all these uses. We will be confined to some typical applications of tracers in medicine. The applications of tracer methods in the field of medicines can broadly be divided into two groups :

(i) The use of tracers (in microcurie closes) in diagnostic methods for localizing bodily disorders; and

(ii) The use of tracers (in large doses) for therapeutic purposes in the treatment of certain abnormal conditions in the body,

Now, the question is : Why is the classification of tracers into two groups? it seems clear that certain radioisotopes are preferentially absorbed by definite types of tissues in the body. The different tissues may be susceptible to different disorders. Hence, for monitoring certain processes in the body and for destroying abnormal disorder in the tissues, is the classification.

Radioiodoin

I^{131} is helpful in detecting disorders of thyroid gland and may cure some of such disorderss. Radioactive iodine and certain other labeled atoms are preferentially adsorbed by cancerous cells. This fact has been made use of in locating brain tumours and sometimes their limits of growth.

Radiosodium

Na^{24} has been used for examining circulation of blood. A small known amount of sodium chloride solution containing Na^{24} is injected into the patient's arm and the time required for its arrival various other parts of the body, as detected by a counter, is an indication of the normal or impaired circulation of blood. Timings observed are compared with the standard data for a normal person. Any local obstruction in any part of the body can be indicated by slowing down of circulation over that part, and treatment can be made accordingly. As described earlier, radio sodium can be used to assess the volume of blood in a patient of anaemia.

Radioactive Phosphorous

P^{32} has been used for locating bone fractures in the patients. It is a fact that the fast growing cells tend to concentrate phosphorus more than the normal cells. This fact has been made use of in locating some forms of cancer and malignant growths in a patient.

Molecules labeled with P^{32} or S^{35} can be used to study the relationship between metabolism in the brain and the level of its functional development. P^{32} has also been used to measure the number of red cells circulating in various parts, of the body.

Radioactive Iron

Fe^{59} has been used to examine the disorders associated with pregnancy. It is used to improve methods for storing blood for transfusions. Gamma radiations from radium have long been used for the treatment of cancer. Since Co^{60} is a gamma emitter, it is replacing the use of radium for curing cancer. It is cheaper and safer to use. The gamma radiations are also replacing the use of X-rays in making X-ray pictures. Gold-198 has also been used in curing some forms of cancer. The use of radioisotopes in the study of absorption, metabolism and in knowing the safety limits of toxic drugs has made outstanding contributions in medical research.

Some important medical applications of radioactive tracers are given as follows :

(a) Radiation has long been used in medical therapy as a means for controlling the development and growth of cells, *e.g.*, in the treatment of some forms of cancer. The penetrating power of X-rays, produced by X-ray mechines, is utilized in teletheraphy to irradiate abnormal tissue deep inside the body. Since about 1988, penetrating beams of high-energy protons and alpha particles have also been used effectively for the same purpose. In another type of radiation treatment, small capsules or needles containing radium (or its decay products) are implanted within a particular organ which is thus subjected to the action of gamma rays. At the present time, artificial radioisotopes provide more convenient and often cheaper alternative sources of radiation both in teletherapy and for internal placement.

(b) Radioisotope teletherapy units most of which use cobalt-60 as the source of gamma rays (1.17-and 1.33-MeV energy), have been installed in hospitals in many parts of the world. The cobalt-

60 is made by exposing normal cobalt metal, consisting exclusively of cobalt-59 to the action of slow neutrons in a nuclear reactor; the cobalt-60 is then readily formed by the $_{59}$Co (n, γ) ^{60}Co reaction. The chief drawback of cobalt-60 as a radiation source is its relatively short half-life of 5.26 years; this means that the cobalt must be removed from the teletherapy unit every few years and reexposed to neutronss. As an alternative, cesium-137; extracted from fission products, is employed as the gamma-ray source in some cases. Cesium-137 has a half-life of 30 years, but the gamma-rays, which are mostly of 0.66-MeV from cobalt-60.

(c) Small cylinders (or needles) made of cobalt-60 of high specific activity encased in silver or gold are used as body implants, thereby serving as cheap substitutes for radium. Other internal irradiation sources are metallic gold-198 and yttrium-90 in the form of ceramic beads of the sesquioxide, Y_2O_3. These substances are not affected by body fluids and so do not require encapsulation. Yttrium-90 differs from the other radioisotope sources in the respect that it emits beta particles but not gamma rays; since the range of the beta particles in tissue is relatively small, the radiation effect is restricted to a limited region in the vicinity of the implant.

(d) Radioisotopes have been used in another manner for internal irradiation, by taking advantage of the preferred absorption of certain elements in particular organs or tisues of the body. Since 1941, iodine-131, which is rapidly taken up by the thyroid gland, has been applied extensively in the treatment of hyperthyrodism, a condition of enlargement and overactivity of the thyroid. Another isotope, iodine-132, is sometimes preferred because it has a shorter half-life, namely, 2.33 hours, compared with 8.06 days for iodine-131; since iodine-132 decays more rapidly it represents less of a hazard when that iodine is released from the thyroid gland into the blood stream.

Although, radioiodine has attracted much interest, the first radioactive species to be employed in radiation therapy. This isotope, in the form of a solution of sodium phosphate in water, is used in the treatment of some blood abnormalities, *e.g.*, chronic leukemia and, particularly, polycythemia. The radiophosphorus is taken up by the bone marrow where the red blood cells and some of the white cells are produced; the beta particles emitted by the phosphorus-32 then inhibit the excessive formation of these cells.

(e) An internal source of alpha particles (and lithium ions) for the treatment of brain tumors is obtained indirectly by utilizing the (n, α) eaction between slow neutrons and stable (nonradioactive) boron-10. A suitable boron compound is injected into a vein of the patient; the brain tumor absorbs the boron from the blood stream but the normal brain cells do not.

A few minutes later, the head is exposed to thermal neutrons from a reactor for about 20 minutes. The neutrons penetrate the head and interact with the boron-10 in the tumor. The alpha particles and lithium ions formed cannot .travel very far and they expend all their energy within a small volume, thereby causing destruction of the abnormal cells in the tumor. The other parts of the brain remain essentially unaffected.

(f) An interesting scheme for achieving what is effectively internal radiation therapy by means of an external source so the extracorporeal irradiation of blood. The idea was apparently first conceived by J. F. Heymans in France in 1921, but it is only since 1962 that it has been developed with some success at the Brookhaven National Laboratory. In certain types of leukemia, there is an overproduction of the white blood cells (lymphocytes) formed in the lymphoid tissues; these cells, however, are readily destroyed by radiation whereas most other cells, are relatively resisant. A loop of plastic tubing is connected outside the body between an artery and a vein, *e.g.*, in the arm of a human subject, and blood is circulated tlirough the tubing by the natural pumping action of the heart. Part of the plastic loop is wound around d strong gamma-ray source, consisting of cobalt-60 or cesium-137, in a shielded container. As the blood passes by the source, it is exposed to gamma rays and the number of lymphocytes is decreased. The treatment, which lasts a few hours, is repeated as required.

Diagnostic Applications of Radioisotopes in Medicine

Radioactive isotopes have become a very valuable tool in the diagnosis and understanding of many diseases. They have proved to be particularly useful in certain cases where conventional diagnostic procedures have not been too satisfactory. A few examples of the many applications of radioisotopes as diagnostic tools will be given here. A simple instance is provided by the use of radiosodium to study cases of resricted circulation of the blood. A small quantity of sodium chloride solution, in which the sodium has been labeled with sodium-24, is injected into a vein of the patient's forearm; a gamma-ray counter is then placed in contact with one of the feet.

If the blood circulation is normal, the presence of radioativity is very soon detected in the food; it increases rapidly and reaches a maximum value within less than an hour. If there is a circulatory impairment of some kind, however, the radioactivity will increase slowly, showing that the blood has difficulty in reaching the foot. By moving the counter to different parts of the body the postition of the restriction can be located and the necessary treatment can be applied.

A modification of this scheme has been proposed for studying the pumping action of the heart. Labeled sodium chloride is injected into the blood stream, as before, and a counter, attached to an automatic pen-recorer, is placed over the heart. As the radio sodium enters the right side of the heart the pen of the recorder rises and then drops as the venous blood enters the lungs. A few seconds later, the radiosodium appears with the arterial blood on the left side of the heart and there is another rise and fall of the recording pen. By examining the resulting curves the pumping action of the two sides of heart can be compared and abnormalities can be discovered. Another substance which has been utilized extensively in examining the action of the heart is albumen from human blood serum labeled with radioactive iodine-131. This is injected into a vein and the gamma-ray activity observed with a counter locaed close to the heart.

It has been found by introducing iron-59 into the blood of a normal human being that the iron is deposited in and then removed from the marrow, the spleen, and the liver in a manner, that is specific for the particular organ. In cases of anemia (deficiency of red blood cells) and of polycythemia (excess of red cells), the normal behaviour is changed in a manner typical for each disease. In refractory anemia associated with hypoplastic (underdeveloped) marrow, for example, the iron is found preferably in the liver rather than in the marrow where the red blood cells are formed.

By using the positron emitter iron-52 as a tracer, in conjunction with a positron camera, it has been shown that the blood-forming morrow is normally present in the spine with very little in the bones of the arms and legs. In certain blood diseases associated with excessive red cell formation, however, the active marrow is also found in the arms and legs. On the other hand, in some cases of severe anemia the marrow that produces the red blood cells leaves the skeleton and migrates to the spleen where it is less effective. The tracer experiments have also demonstrated that red cell formation is controlled by a hormone, called erythropoietin, which circulates in the bloodstream. Abnormal conditions are associated with an excess or deficiency of this hormone.

Measurements by means of whole-body counters of the total amount of iron-59 retained by the body have confirmed the conclusions, reached from other observations that in a normal person the absorbed iron is eliminated very slowly. On the average, the daily loss of iron from a healthy person is equivalent to that in about one cubic centimeter of blood; this loss occurs mainly through the walls of the intestine. In persons suffering from various diseases of the blood or of the gastrointestinal tract the rate of loss of iron, as determined by the whole-body counter, exceeds the normal rate in a manner that is typical of each disease. The whole-body counter can thus serve as a diagnostic tool.

Vitamin B_{12} is a complex organic compound containing the element cobalt. It is normally stored in the liver and is released into the blood stream as required. The viamin has been prepared with the cobalt as a radioactive isotope, either cobalt-57, -58, or -60,and its rate of absorption and loss from the body has been studied with the whole-body counter. From the results it is possible to distinguish three types of behaviour: first, normal; second, inability to absorb vitamin Bu from the intestine; and third, pernicious anemia in which there is a deficiency of the vitamin in the blood. Whole-body radiation measurements, with labeled vitamin Biz.are being used in the study of pernicious anemia.

An interesting application of whole-body counters is based on the natural radioactive potassium-40 that is always present in the body. This isotope decays with the emission of a beta particle and a gamma-ray, but the rate of decay is so relatively slow that it requires a whole-body counter to detect the radiation.

Potassium is found in nearly all parts of the body except the fatty tissue;consequently, a measurement of the total quantity of potassium-40, together with other data, makes it possible to determine the relative proportions of lean and fatty tissue. Such information can be used to study the changes in the body composition that take place throughout life, both in normal circumstances and as a result of various diseases, and their relation to the diet. In addition, correlaions can be made with other physiological measurements, such as the basal metabolic rate. It is of interest to note that in the average adult man 20 percent of the body tissue is fat, whereas in a woman the average is 26 percent. Exceptionally lean people may contain as little as 2 percent of fat; at the other extreme, those suffering from obesity can have as much as 40 percent of fatty tissue.

The fact that certain elements tend to concentrate to different extents in normal and abnormal tissue has been utilized to detect the presence and identify the position of tumors with the aid of radioisotopes. This procedure serves as a means of diagnosis and also facilitates treatment of the abnormal condition by surgery or radiotherapy. The element iodine is rapidly absorbed by the thyroid gland where it is stored as the compound thyroxin.

The function of the gland can thus be readily studied by giving the subject an aqueous solution containing a small quantity of sodium iodide containing radioactive iodine-131 and then observing the gamma-ray emission from the thyroid gland. It has been found, by utilizing a scanner or a scintillation camera, that abnormal regions can be detected by their failure to absorb iodine. Such regions sometimes contain malignancies, but cancer rarely occurs in the portions of the thyroid gland that take up iodine in a normal manner.

Brain tumors are often difficult to detect and especially to locate. Several different substances labeled with radioisotopes have been found useful in this connection because they tend to concentrate in the tumor. One of the first materials employed was serum albumin combined with iodine-131, but better results have been obtained more recently with neohydrin labeled with mercury-203, with the positron emitter gallium-68 chelated with ethylenediamine tetraacetic acid and with the isomer of technetium-99 in the form of the pertechnetate ion. After injection of the radioactive tracer, the location of the brain tumor can be found by means of a scintillation camera or a multiple detector scanner.

The functioning of various organs, such as the liver, kidney, and spleen, under normal and disased conditions, and the identification of abnormal areas have been studied by means of suitable radioactive isotopes. The rates of absorption and loss of the isotope, after introduction into the body, can be observed by a gamma-ray counter placed adjacent to the particular organ, whereas abnormal regions are detected by a scanner or camera. For liver studies the common tracer is the dye rose bengal labeled with iodine-131. for the kidney the tracer is hippuran also labeled with iodine-131, whereas for the spleen use is made of red blood cells labeled with chromium-51. For scanning the pancreas, seleno-methionine, containing radioactive selenium-75, has been developed. Methionine, an essential amino acid containing sulphur, is taken up rapidly by the pancreas and the same is true for the modification in which the sulphur atom in the molecule is replaced by the chemically analogous selenium.

SOME SOLVED PROBLEMS

Problem 1:

It is found 46.3 *mg of naturally occurring potassium show a b-activity of* 1.5 *dis/sec. The isotope responsible for this activity is* K^{40} *which makes up* 0.012% *of the natural mixture. Calculate the half-life of* K^{40}.

Solution:

The number of K^{40} atoms in 46.3 mg of natural potassium is

$$N = \frac{0.012}{100} \times 46.3 \times 10^{-3} \text{ gm} \times \frac{6.023 \times 10^{23} \text{ atoms/mole}}{40 \text{ grams/mole}}$$

$$= 2.37 \times 10^{10} \text{ atoms}$$

$$T_{1/2} = \frac{0.693}{\lambda}$$

But $$\frac{-dN}{dt} = N\lambda$$

or $$\lambda = \frac{-dN}{dt}/N$$

$$\therefore \quad T_{1/2} = \frac{0.693\,N}{\frac{-dN}{dt}}$$

$$= \frac{0.693 \times 2.37 \times 10^{16} \text{ atoms}}{1.5 \text{ dis./sec}}$$

$$= 5.8 \times 10^{16} \text{ sec.} = 1.23 \times 10^{6} \text{ years.}$$

Problem 2:

A sample of carbon from a wood recovered in excavation is found to give 700 counts per minute per gram of carbon. What is the approximate age of the of wood ?

Solution:

The half-life of C^{14} is 5770 year. Therefore,

$$l = \frac{0.693}{5770} = 1.20 \times 10^{-4} \text{ per year}$$

The C^{14} from wood recently cuts down decays at the rate of 15.3 disintegrations per minute per gram of carbon. Therefore.

$$\log \frac{N_0}{N} = \frac{\lambda t}{2.303}$$

$$\log \frac{15.3}{700} = \frac{1.20 \times 10^{-4}\, t}{2.303} \quad \text{or } t = 6520 \text{ years. } \textbf{Ans.}$$

Problem 3:

At radioactive equilibrium, the ratio between atoms of the two radioactive elements. A and B was found to be 3.1 × 109 : 1 respectively. If the half-life period of element A is 2.0 × 1010 years, what is the half-life period of element B?

Solution:

We know that at radioactive equilibrium between element A and B.

$$N_A \lambda_A = N_B \lambda_B$$

or $$N_A \frac{0.693}{(T_{1/2})_A} = N_B \frac{0.693}{(T_{1/2})_B}$$

or $$\frac{N_A}{(T_{(1/2})_A} = \frac{N_B}{(T_{(1/2})_B}$$

or $$\frac{N_B}{N_A} = \frac{(T_{(1/2})_B}{(T_{(1/2})_A}$$

Here $N_B/N_A = 1 : 3.1 \times 10^9$ and $(T_{1/2})_A = 2 \times 10^{10}$ years

$\therefore$ $$(T_{1/2})B = \frac{1}{3.1 \times 10^9} \times 2 \times 10^{10} \text{ years} = 6.45 \text{ years. } \textbf{Ans.}$$

Hence, the half-life period of element B is 6.45 years.

Problem 4:

Calculate the barrier height for an a-particle inside the $_{83}Ra^{226}$ *nucleus.*

Solution:

$$\text{Barrier height} = \frac{2(Z-2)e^2}{R_0\, A^{1/3}}$$

$$= \frac{2 \times 86 \times (4.8 \times 10^{-10})^2}{1.3 \times 10^{-13} \times (226)^{1/3}}\ \text{erg}$$

$$\times \frac{1}{1.6 \times 10^{-6}\ \text{erg/Mev}}$$

$$= 32\ \text{Mev.}$$ **Ans.**

Problem 5:

A thin sample of gold was irradiated in a thermal neutron flux of 10^{12} *neutrons* $cm^2\ sec^{-1}$ *for 25.66 hours. In the reaction the nuclide* Au^{198} *is produced with a half-life of 64 hours. If the thermal neutron absorption cross section is 98 brans, what is the specific activity of the sample ?*

Solution:

$$A = N\,\phi\,\sigma$$

Suppose 1 gm of the gold was irradiated,

Then, $$N = \frac{1}{198} \times 6.023 \times 10^{23}\ \text{atoms}$$

$$\phi = 10^{12}\ \text{neutrons cm}^{-2}\ \text{sec}^{-1}$$

$$\sigma = 98\ \text{barns} = 98 \times 10^{-24}\ \text{cm}^2 \quad [\because 1\ \text{barn} = 10^{-24}\ \text{cm}^2]$$

$$T_{1/2} = 64\ \text{hours and } t = 25.6\ \text{hours}$$

Substituting these values in equation (1), we get

$$A = \frac{6.023 \times 10^{23}}{198} \times 10^{12} \times 98$$

$$\times 10^{-24}\left(1 - e^{\frac{-0.693}{64} \times 25.5}\right)$$

$$= \frac{6.023 \times 10^{23} \times 10^{12} \times 98 \times 10^{24}}{198 \times 3.7 \times 10^{10}}$$

$$\left(1 - e^{\frac{-0.693}{64} \times 25.6}\right) \text{ curies/gm}$$

$$= 1.96 \text{ curies/gm. } \textbf{Ans.}$$

Problem 6:

When 0.6 *gm of a standard sample of* La_2O_3 *was irradiated, an activity of* 450 *c.p.m. gm*$^{-1}$ *was produced. Later, an unknown sample of* La_2O_3 *was irradiated under the same conditions, an activity of* 30,000 *c.p.s mg*$^{-1}$ *was produced. Find the weight of* La_2O_3.

Solution:

It is an example of a comparator method of activation analysis . From this method, we know.

$$\frac{\text{Wt. of } La_2O_3 \text{ in unknown}}{\text{Wt. of } La_2O_3 \text{ in s tan dard}}$$

$$= \frac{\text{Total activity from } La_2O_3 \text{ in unknown}}{\text{Total activity from } La_2O_3 \text{ in s tan dard}}$$

or $$\frac{\text{Wt. of } La_2O_3 \text{ in unknown}}{0.6 \text{ gm}} = \frac{30,000 \text{ c.p.s mgg}^{-1}}{450 \text{ c.p.m.gm}^{-1}}$$

$$= \frac{30,000 \times 60 \times 1000 \text{ c.p.m. mg}^{-1}}{450 \text{ c.p.m.gm}^{-1}}$$

or $$\text{Wt. of } La_2O_3 \text{ in unknown} = \frac{30,000 \times 60 \times 1000}{450} \times 0.6 \text{ g}$$

$$= 3.4 \times 10^6 \text{ g} \quad \textbf{Ans.}$$

Problem 7:

To a protein hydrolysate was added 10.1 *mg of labelled alanine of specific activity 128 c.m.p. mg*$^{-1}$ *measured in a particular counting arrangement. A sample of pure alanine isolated from the mixture was found to have a specific activity of 68.3 c.p.m. mg*$^{-1}$ *when measured in the same counting arrangement. What was weight of alanine present in the hydrolysate ?*

Solution:

From the given data.

$$m' = 10.1 mg$$

$$S' = 128 \text{ c.p.m. } mg^{-1}$$

$$S = 68.3 \text{ c.p.m. } mg^{-1}$$

On substitution of these value in the equation, we get

$$m = m'\left[\frac{S'}{S} - 1\right] mg$$

$$= 10.1\left[\frac{128}{68.3} - 1\right] = 8.83 \text{ mg.}$$

The weight of alanine in the hydrolysate

$$= 8.83 \text{ mg.}$$ **Ans.**

4

Models and Properties of Nuclear

INTRODUCTION

Recent researches, especially in the last few years have provided considerable informations about the nucleus. Some of the important properties of the nucleus are as follows :

CLASSIFICATION OF NUCLEAR

Isotopes are nuclei with the same atomic number Z but different mass number A. (Isotopes-same number of protons). Isotopes of an element have identical chemical behaviour and differ only in mass. Nuclei with the same mass number but different atomic number are called isobars. They have different physical and chemical properties. Nuclei with an equal number of neutrons are called isotones. (Isotones-same number of neutrons). Isomeric nuclei are isomers which are nuclei with same Z and same A but differ from one another in their nuclear energy states. These nuclei are distinguished by their different life times.

Mirror nuclei have equal mass number but their atomic number differ by one' (*i.e.,*) the number differ protons in one equals to the number of neutrons in the other; For example, $_7N^{15}$ and $_8O^{15}$ are mirror nuclei.

NUCLEAR SIZE

The nucleus is about 1000 times smaller than that of an atom and has a mean radius of the order of 10^{-14}m to 10^{-15}m. The empirical formula for the nuclear radius is

$$R = r_0A^{1/3}$$

Where A is the mass number and r_0 is a constant and

$$r_0 = 1.3 \times 10^{-15} \text{ m} = 1.3 \text{ Fermi.}$$

Nuclear mass. If m_p *and* m_n are the protons and neutrons masses respectively then the assumed nuclear mass should be $Zm_p + N_n$.

Accurate experimental determinations by mass spectrometers show the real nuclear mass is less than $Zm_p + Nm_n$.

The difference in mass is called the mass defect," (ie) $Zm_p + Nm_n$- real nuclear mass = Δm

Nuclear density. If the ratio of nuclear mass to nuclear volume is calculated, it works out to be 1.816×10^{17} kgm^{-3}, which shows that the nuclear matter is in an extremely compressed state.

Existence of electrons inside the nucleus. For the electron residing inside a nucleus the uncertainty in its position may not exceed 10^{-14} m .since nuclei are less than 10^{-14} m in radius. Using Heisenberg's uncertainty principle the uncertainty in electron momentum would be

$$\Delta p \ {}^{3} \frac{k}{\Delta x} = \frac{1.054 \times 10^{-34}}{10^{-14}} \geq 1.1 \times 10^{-20} \text{ kgm/s}$$

Assuming the momentum p of electrons to be the same as Δp, the total energy of the electron is given by

$$E^2 = p^2c^2 + m_0^2c^4 = 9 \times 10^{-24} + 6.6. \times 10^{-27}$$

or $$E \sim 3 \times 10^{-12} \text{ joules} = 20 \text{ Mev.}$$

(*i.e.*,) free electrons confined with in the nucleus would have a K.E. of the order of 20 Mev. But experimentally electrons emitted by radio - active nuclei have never been found to have kinetic energies greater than about 4 MeV. This large discrepancy indicates that nuclei cannot contain free electrons.

ELECTRICAL AND MAGNETIC PROPERTIES OF THE NUCLEUS

These properties explain various nucleus phenomena. Each neutron and proton is associated with angular momentum. These particles spin around an axis passing through their centres of masses. This spinning motion gives rise to spin–angular momentum (S) whose magnitude is 1/2. $\frac{h}{2\pi}$. According to wave–mechanical concept, spin angular momentum is having two orientations, one parallel and the other anti-parallel. The orbital angular momentum (L) is a vectorial quantity whose maximum possible component would be an integral multiple of $\frac{h}{2\pi}$. Thus,

Total angular momentum (I) = L ± S

The magnetic moment (μ) of the nucleus would be represented by the following equation

$$\mu = \gamma . \frac{h}{2\pi} . I = g_N \, \mu_N . I$$

Where γ denotes nuclear gyromagnetic ratio, g_N, the nuclear Lande's splitting factor and μ_N, the nuclear magneton (5.04929 × 10^{24} erg/ gauss). The g_N can be expressed as

$$g_N = 1 + \frac{I(I+1) - L(L+1) + S(s+1)}{2(I+1)}.$$

Where S denotes resultant spin number (Σs_i) and L, resultant orbital number (Σs_i) and I = L ± S.

The most useful method for determining magnetic moment is nuclear magnetic resonance (NMR). Nuclear spin and magnetic moment are quite useful in understanding the complexities of nuclear structure. If the number of protons and neutrons is even, I = 0, The magnetic moment of such a nucleus would also be zero.

Experiments have also revealed that such nuclei have zero magnetic moment. Electrical quadropole moment, Q, is related to the shape of the nucleus. This quantity is regarded as a measure of the deviation of the nucleus from spherical symmetry. It is measured as *eQ*, where *Q* is the measure of deviation from spherical symmetry which may be ≥ 1.

If the nucleus is an ellipsoid having diameter as 2a along the symmetry axis and 2b perpendicular to this axis and if the charge is uniformally distributed throughout the volume of the ellipsoid, then *Q* may be put as follows :

ELECTRIC QUADROPOLE MOMENT

Due the symmetry of nuclei about the centre of mass, in stationary states, for atoms and nuclei, the electric dipole moment is zero. A deviation from the spherical symmetry can be expressed in terms of electric quadropole moment. Since most of the nuclei assume the shape of an ellipsoid of rotation, they have an electric quadropole moment. The dimensions of quadropole moment is that of an area and in nuclear physics the unit used is a Barn (1 Barn = 10^{-28} m^2).

Spin considerations. As electrons and protons each has spin 1/2 k, the nucleus must have integral spin if it contains an even number and half integral spin if it contains an odd number of particles (protons + electrons) in the nucleus. But experiments show that spin depends on mass number A of the nucleus. If A is even, spin is zero or an integer. If A is odd, it is an half integral. For an atom $_7N^{14}$ the number of particles inside the nucleus will be 2A–Z. Hence for such nucleus the spin must be half integral value, which is against observed facts.

Magnetic Moment Consideration. Since the magnetic moments of a proton is nearly 1837 times less than that of an electron, the magnetic moment of electrons will have a dominant influence on the nuclear magnetic moments. But the magnetic moments of all nuclei are small compared to the magnetic moment of electrons. These shows that electron is not a constituent of the nucleus.

Compton Wavelength. Several theoretical considerations suggest that a bound fundamental particle cannot be localised in the region smaller than its compton wavelength h/m_c. For an electron the compton wavelength works out to be $150 \times 10^{-14} \sim 250$ nucleon diameter. This excludes the possibility of finding an electron inside a nucleus.

Parity Properties of Nuclei - If spatial coordinates (X, Y, Z) are replaced by (X, – Y, – Z) and the spatial part of its wave function does not change, the motion of the nucleus is said to possess even parity. This transformation of coordinates is equivalent to reflection of the nucleus position about the origin of X, Y, Z system of axes. On the other hand if the transformation of coordinates causes a change of sign of spatial part of the wave function, the nucleus is said to have odd parity. Thus, the parity of the nucleus in a given state depends upon the value of the orbital angular momentum (L). The parity would be said to be odd if L is odd and even if L is even.

Statistical Properties of Nuclei. The classical Maxwell-Boltzman statistics is only successful in explaining velocity and energy distribution of molecules in gases, but it fails to explain the combined properties of proton-electron, neutron-proton and neutron. On the basis of quantum mechanics, two statistical theories have been developed, viz. Bose-Einstein and Fermi–dirac statistics.

BOSE-EINSTEIN STATISTICS

It is applicable to the nuclei having even mass numbers. Thus, it may be concluded that the total angular momentum of the nuclei conforming to Fermi-Dirac statistics are odd half integral multiple of $\frac{h}{2\pi}\left(\frac{1}{2}\cdot\frac{h}{2\pi},\frac{3}{2}\cdot\frac{h}{2\pi},\frac{5}{2}\cdot\frac{h}{2\pi}...\right)$ and of those following Bose-Einstein statistics are integral multiple of $\frac{h}{2\pi}\left(\frac{h}{2\pi},\frac{2h}{2\pi}.\frac{3h}{2\pi}...\text{etc.}\right)$.

Fermi - Dirac statistics is applicable to such systems for which the wave function is anti-symmetrical, implying thereby that spin of the wave function changes when all the coordinates of the two identical particles (three spatial and one spin) are exchanged.

On the basis of characteristic of the wave equation, Pauli's exclusion principle is applicable to all those particles which are conforming to Fermi-Dirac statistic. It is experimentally verified that like nuclei of odd mass number (A), electrons are obeying this statistical theory.

ATOMIC MASS UNIT

On the physical scale, it has been agreed internationally to use a unit called an atomic mass unit (a.m.u.), based on the mass of an atom of the carbon isotope C^{12}. By definition.

12 atomic mass unit (a.m.u) = mass of carbon atom C^{12}.

$\therefore$ 1 a. m. u. = 1/2 th of mass of carbon atom C^{12}.

Now 6.02×10^{23} is the number of atoms in 12 grams of the isotope, C^{12}.

$$1 \text{ a.m.u.} = \frac{1}{12}\times\frac{12}{6.02\times10^{23}} \text{ gram.}$$

$$= 1.66 \times 10^{-24} \text{ g or } 1.66 \times 10^{-27} \text{ kg.}$$

We know that 1 gram change of mass produces an energy change of 9×10^{20} ergs and that

$$1\text{MeV} = 1.66 \times 10^{-24}\ ergs$$

$$1 \text{ a.m.u.} = \frac{1.66\times10^{-24}\times9\times10^{20}}{1.6\times10^{-6}}\ Mev$$

$$1 \text{ a.m.u.} = 931\ MeV.$$

This relation is used to change mass unit to MeV and vice versa.

MASS DEFECT

Aston's precision measurements revealed that in practically every case, small deviations from the whole number rule existed. Typical exact weights being, for example,

$$H = 1.00813 \text{ a.m.u.}$$

$$^{4}He = 4.0038 \text{ a.m.u.}$$

$$^{40}Ca = 39.996 \text{ a.m.u.}$$

The importance to be attached to these deviations becomes apparent when it is realised that the helium nucleus consists of two neutrons and two protons. The helium atom should apparently have a mass of two hydrogen atoms plus two neutrons. The mass of hydrogen atom is 1.00813 and that of neutron is 1.00898 a.m.u. So that mass of the helium atom should be = 2 (1.00813 + 1.00898) = 4.03422 a.m.u. Helium atom is in fact lighter by about 0.03 a.m.u.

This difference can be explained by the theory of relativity, according to which mass and energy are interchangeable and can be converted into one another. Inside the helium nucleus the constituent panicles are very tightly packed together. It can be shown that in the original process of the formation of such a nucleus energy must be released in very large amount before a stable packing state is reached. The enormous loss of energy which is related to the binding forces between the particles means a corresponding loss of mass. It means that the stability of nucleus depends upon the amount of mass which has been lost in the form of energy.

The deviation of the mass of nucleus from the whole number is called the mass defect. Mass defect is also defined as : where M is the atomic mass and A is the mass number. A is equal to the sum of the number of protons and neutrons in the nucleus. Mass defect represents the binding energy of the nucleus. Mass defect may also be defined as the amount of mass which Would be converted into energy if a particular atom has to be assembled from *its* constituents. The energy equivalent of mass defect is a measure of binding energy of the nucleus.

The, difference in the atomic mass of an isotope and- mass .number value (mass defect) was, expressed by F. W. Aston (1927) as packing fraction by the following expression :

$$f = \text{Packing fraction} = \frac{\text{Isotopic mass} - \text{Mass number}}{\text{Mass number}} = \frac{\Delta}{A}$$

Thus, the mass of one of the chlorine isotopes, Cl^{35} is 34.980. The packing fraction, therefore, is

$$f = \frac{34.980 - 35}{35} = -\ 0.00057$$

The usual way of expressing packing fraction is parts per. 10000. So the packing fraction of Cl^{35} isotope, according to the convention, is– $0.00057 \times 10^4 = -\ 5.7$.

The isotopic mass numbers are expressed on the so called *physical scale* which is based on O^{16} isotope of oxygen being exactly 16. Hence, the packing fraction of O^{16}, by definition, is zero.

Packing fraction is the 'mass -defect per nucleon. It is a very important quantity and is a measure of the stability of the nucleus. It varies from one nucleus to other. This is sometimes positive or zero but more often negative.

SIGNIFICANCE OF PACKING FRACTION

(i) A negative packing fraction means that the atomic mass is less than the nearest whole number and suggests that there has been & conversion of mass into energy in the formation of the particular nucleus. Since this same amount of energy would have to be supplied in order to break up the nucleus, it appears that a negative fraction implies exceptional nuclear stability.

(ii) A positive packing fraction generally indicates instability of the nucleus. This statement is not strictly correct especially for small atomic mass elements as the mass of both the protons and the neutrons are slightly higher than unity.

In general, the lower the packing fraction of an element, the greater the stability of its nucleus.

When the observed packing fractions are plotted against the mass numbers of the atoms, they are found to fall upon the smooth from the curve it follows that :

1. It is seen that with the exception of He^4, C^{12} and O^{16} nearly all elements fall on or very close to the curve. The relatively low packing fraction indicates exceptional stability of the elements.

2. The lowest values of packing fraction are observed for the transition elements of iron family which indicate maximum stability of their nuclei. From here on, the values again rise and with mass numbers higher than 200 they become positive, showing increasing stability of the nuclei. This is proved 'by the phenomenon of radioactivity exhibited by high atomic weight elements. The nuclei in such case gradually disintegrate, ultimately, giving a lighter and, more stable nucleus (usually an isotope of lead).
3. In the intermediate range of mass numbers, the packing fractions are negative, showing that the nuclei are less stable.

NUCLEAR ISOMERISM

It has been observed that in various cases of artificial radioelements, *one and the same nucleus with same mass number and atomic number,* disintegrates in different ways with different decay periods. This difference in radioactive properties is due to difference in internal structure of nuclei which are otherwise similar in all respects. These nuclides are called **nuclear isomers** and the phenomenon as **nuclear isomerism** by analogy with chemical isomerism.

In 1917, **Soddy** suggested that such isomers might exist among the natural radio-elements. **Hahn** in 1921 established experimentally the isomeric property of UX_2 and UZ. In recent years about 17 *isomeric pairs* have been found among artificially prepared nuclides.

The existence of different energy levels of the nucleus of an a tom represents different isomeric states. In the case of nuclear isomers, the difference in the energy between the excited state and ground state is comparatively small while the nuclear spin difference is considerably large, with the result that the transition from upper to the lower level will be forbidden. Therefore, the excited state will exist for a measurable, through small period. It is in these cases the phenomenon of nuclear isomerism is observed. These nuclides are isotopic isobars or isobaric isotopes having same decay scheme with different halflives.

Example. 1 A well know example is that Zn^{68} obtained from Zn^{69} by (d,p) reaction.

$$_{30}Zn^{68} + {}_{1}D^{2} \rightarrow {}_{30}Zn^{69} + {}_{1}H^{1}$$

13.8h $\uparrow$ (excited) $\rightarrow$ The excited state decays with emitting γ-rays

$$_{31}Ga^{69} \xrightarrow[75\,Min.]{-\beta} {}_{30}\dot{Zn}^{69} \text{ (ground state)}$$
(stable)

Here Zn^{69} shows nuclear isomerism.

2. *The isomerisms of* Br^{80}. It was the first case to be discovered among artificial ratio–elements.

There are only two known stable isotopes of bromine of mass numbers 79 and 81. When bromine is bombarded by slow neutrons, radioactive elements are formed which disintegrate by the emission of electrons with three different periods, 18 mins., 4.2 hrs, and 36 hrs. Chemical tests show that the radioactive products are isotopes of bromine.

$$\left.\begin{array}{l} _{35}Br^{79} + {}_0n^1 \rightarrow {}_{35}Br^{80} + \gamma \\ _{35}Br^{80} \quad\quad\quad \rightarrow {}_{36}Kr^{80} + {}_1e^0 \end{array}\right\} \text{For mass 79}$$

and

$$\left.\begin{array}{l} _{35}Br^{81} + {}_0n^1 \rightarrow {}_{35}Br^{82} + \gamma \\ _{35}Br^{82} \quad\quad\quad \rightarrow {}_{36}Kr^{82} + {}_{-1}e^0 \end{array}\right\} \text{For mass 81}$$

Br^{80} are Br^{82} are therefore the only two radioactive isotopes responsible for the three observed periods which means that one of these will have a double period. The identity of this isotope was determined by bombarding bromine with γ-rays of very high energy (17 MeV) when radioactive isotopes Br^{78} and Br^{80} are obtained. These isotopes decay emitting β-rays again with three different periods viz., 6.3 mins., 18 mins., 4.4 hrs.

$$\left.\begin{array}{l} _{35}Br^{79} + \gamma \rightarrow {}_{35}Br^{78} + {}_0n^1 \\ _{35}Br^{78} \quad\quad \rightarrow {}_{34}Br^{80} + {}_{+1}e^0 \end{array}\right\} \text{For mass 79}$$

$$\left.\begin{array}{l} _{35}Br^{81} + \gamma \rightarrow {}_{35}Br^{80} + {}_0n^1 \\ _{35}Br^{80} \quad\quad \rightarrow {}_{36}Kr^{80} + {}_{-1}e^0 \end{array}\right\} \text{For mass 79}$$

As Br80 was formed in both the cases so it is the isomeric nuclei with 11 minuted and 4.4 hours half lives.

$$\underset{\text{Excited}}{_{35}Br^{80}} \xrightarrow[4.4\text{ hrs.}]{\gamma\text{–rays}} \underset{\text{Ground state}}{_{35}Br^{80}} \xrightarrow[18\text{ mins.}]{-} \underset{\text{stable}}{_{36}Kr^{80}}$$

Other examples are

Mn^{52}, Co^{55}, Rh^{104}, $Te^{127,\ 130,\ 131}$, ‘Ag^{106}, Sr^{80} etc.

Theoretical explanation. Wiezsacker (1936) suggested a theory to ex plain nuclear isomerism. It is based on the existence of low excited states whose *angular momentum* (*i.e.*, nuclear spin) is considerable, differing by several units from that of the ground state. Such an excited state would be *metastable* and nucleus can remain in this state for a sufficient time and it may be regarded as a separate nucleus. The metastable nucleus would have its own decay period and may decay in the following different ways :

(i) The metastable state may go over to the ground state with the emission of γ-rays where upon the g round state will decay further with b-emission.

(ii) The metastable state may emit a β-particle directly, this will take place only if the probability of β-emission is smaller than that of γ-emission for the given metastable state.

Theoretical considerations have shown that sufficient metastable states will not exist for light nuclei. It is because of small density of levels for such nuclei and due to small angular moments. It has been further shown that metastable states are stable nuclei. But they are detected with great difficulty because of the emission of very soft γ-rays only.

Goldhaber and coworkers could classify the nuclear properties of 77 isomers, for which half-life time is between 1 sec and 8 months. About half of these are M_4 transitions and the remainder are M_3, E_3 and E_4. The shorter lived group (life times between 10^{-3} and 10^{-9} sec) is made up of M_1, M_2 and E_2 transitions. The transitions accompanied by E_5 radiations are not common.

Energy level Scheme for Isomeric Nuclei

m = metastable state

g = grcund state

Half period of isomeric transition. Most known γ- decay rates have been determined by the direct measurement of the life - times of the excited states, *i.e.*,

Total decay states,

$$\lambda = \lambda_y + \lambda_c = \lambda_y (1 + \lambda_a)$$

$$T_{1/2} = \frac{\log_e 2}{\lambda_y} = \frac{0.693\, \tau_y}{1+\alpha}$$

As the internal conversion coefficient a can be measured or calculated theoretically and half life $T_{1/2}$ can be measured and hence λ_y the average life or $\lambda_{y'}$ the rate of emission can be calculated.

BINDING ENERGY

When two particles approach each other under the influence of an attractive force, their energy decreases ; the difference being emitted in the form of radiations. Consequently, when nucleons interact attractively and coalesce to form a nucleus, their energy must be less than what is when they were separated.

According to the Einstein mass–energy relation, their combined mass in the nucleus must be less than the sum of their mass before their fusion by an amount corresponding to this loss of energy. Therefore, we come to the conclusion.

"The mass of a nucleus must be smaller than the sum of the mass of its constituents when they were separate".

"Total binding energy of a nucleus is the energy with which all nucleons are held together in that nucleus.

"The binding energy of a nucleus is the energy which must be given to a nucleus to break it completely into its protons and neutrons". Since the mass change is a measure of the energy change, from Einstein's mass energy relation, it follows that, if M_H is the mass of a proton, M_N is the mass of a neutron and M is the mass of nucleus, measured by a mass spectrograph, all in a.m.u., then

B = *Binding energy of a nucleus* $= Z\,M_H + (A - Z)\,M_N - M$ a.m.u.

The binding energy may either be expressed in a.m.u. or converted into MeV as one a.m.u. is equal to 931 MeV.

As an example, consider a helium nucleus $2He^4$. This has four nucleons, 2 protons and 2 neutrons. The mass of a proton is 1.0073 and the mass of a neutrons is 1.0087 a.m.u. Total mass of 2 protons plus 2 neutrons

$= 2 \times 1.0073 + 2 \times 1.0087 = 4.0320$ a.m.u. But the helium nucleus has a mass of 4.0028 a.m.u.

$\therefore$ Binding energy = mass difference of nucleons and nucleus.

$= 4.0320 - 4.0028 = 0.0292$ a.m.u.

$= 0.0292 \times 931\ MeV = 27.2\ MeV.$

The binding energy per nucleon B of a nucleus is the binding energy divided by the total number of nucleons. In the case of helium nucleus, since there are four nucleons (2 protons and 2 neutrons), the binding energy per nucleon is 27.2/4 or 6.8 MeV.

Binding energies vary greatly with the composition of the nucleus. An idea of the relative stability of the stable nuclei of the different elements can be had from a plot of the binding energy per nucleon of the nucleus against the mass number of nucleus. A study of the figure reveals following three interesting points :

(i) The first point is that the curve shows a maximum around the nuclei of mass number 60.

(ii) Secondly, nuclei of very heavy elements like uranium have comparatively low binding energy and will be apt to splitting into smaller nuclei. Such splitting will release great amounts of energy.

(iii) Thirdly, the binding energy per nucleon rises sharply from the isotopes of hydrogen to those of next heavier elements like helium or lithium. Thus fusing together (nuclear fusion) of hydrogen atoms so as to form helium nuclei will provide vast amount of energy. In fact, the energy available form nuclear fusion is far greater than the energy obtained form the fission of an equal mass of a very heavy element.

We know that the binding energy of an electron in the outer part of the atom; outside the nucleus, is of the order of electron volts. Inside many stable nuclei, the binding energy per nucleon is a million times greater. This show that nuclear forces are extremely powerful. They are short–range forces acting over distances $2 - 3 \times 10^{-13}$ *cm* (10^{-13} cm is called 1 Fermi).

These are completely different from electromagnetic forces, which influence electrons outside the nucleus. In the nucleus, the force between a proton and a neutron, or between two protons or two neutrons appears to be the same if allowance is made for the presence of the electric charge.

Packing Fraction and Binding Energy per Nucleon : We have seen that the packing faction f is given by $\frac{M-A}{A}$ where A is the integral mass number and M is the exact atomic mass.

$$f = \frac{M-A}{A} \text{ or } \frac{M}{A} = 1 + f \qquad ...(i)$$

Now $\Delta M = \{ZM + (A + Z) M_N\} - M\}$ is the binding energy 'B". The binding energy per nucleon, or $\bar{B}$ is then given by

$$\bar{B} = \frac{B}{A} = \frac{Z}{A} M + \left(1 - \frac{Z}{A}\right) M_N - \frac{M}{A}$$

$$= \frac{Z}{A} [M - M_N] + M_N - (1 + f)$$

$$= -\frac{Z}{A} [M_N - M] + 0.008665 - f$$

Taking average value of $\frac{Z}{A}$ as 0.45, we get

$$\bar{B} = - 0.00038 + 0.008665 - f = 0.008175 - f,$$

so that the minimum value of f corresponds to maximum value of B^-. Note that since f rarely exceeds 10^{-3} a.m.u the value of $\bar{B}$ is roughly constant. Taking an average value of f as 7 × O^{-4} a.m.u. we find that B = 0.0090 a.m.u. about, or 8.4 MeV for most nuclei (1 a.m.u. = 931 MeV). The reason for this fairly constant value of B is that it is made up largely of the neutron mass excess (M_N –1). This is really a consequence of the fact that all nuclear forces are short range forces.

Semi-empirical Formula of Total Binding Energy. We will now consider the contributions of various terms before deducing the final equation of total binding energy.

(a) Volume energy term. The value of this term is $a_v A$, where A denotes the mass number and a_v a coefficient whose value is determined from known values of mass. This term is also known as *exchange* energy because it originates from exchange forces. This is denoted as BE_1.

(b) Surface energy term. As some of the nucleons of an atom are situated at the surface of the nucleus where they are not surrounded by other nucleons, it means that they are different from the other nucleons present in the bulk of the nucleus. The nucleons in the bulk of the nucleus are surrounded on all sides by other nucleons. Thus the internal exchange forces in the bulk of the nucleus are saturated while those at the surface are unbalanced. The number of the nucleons exposed to the surface may be evaluated in the following way :

The radius of the nucleus,

$$R = R_0 A^{1/3}$$

(Where A denotes the mass number and R_0 the radius of the nucleon)

Surface area of the nucleus = $4\pi R^2 = 4\pi\ R_0^2\ A^{2/3}$

The area of the cross-section of the nucleon whose radius is as follows :

$$R_0 = 4\pi R^2$$

The number of the nucleons exposed to the surface.

$$= \frac{4\pi R_0^{\ 2}.A^{2/3}}{\pi R_0^{\ 2}} = 4\ A^{2/3}$$

$\therefore$ The surface energy term, $BE_2 = -$ as $A^{2/3}$

(Where as is a constant whose value is 13 MeV)

(c) Coulomb's energy term. The coulomb's repulsive forces arise from mutual repulsion between protons. The positive charge of the proton is uniformly distributed throughout the volume of nucleus. This energy term is denoted as BE_3 and makes negative contribution to the total binding energy.

$$BE_3 = -\frac{3}{2.}\frac{Z^2e^2}{R} = -\ \frac{3}{5.}\frac{Z^2e^2}{R_0A^{1/3}}$$

$$= -\ a_c\ \frac{Z^2}{A^{1/3}}$$

Where ac is a constant which equals $\frac{3}{5}\left(\frac{e^2}{R_0}\right)$.

(d) Asymmetry energy term. In heavy nuclei the number of neutrons far exceeds the number of protons. The large excess of neutrons provides the stability to the nucleus against large coulombic repulsive forces which result from large number of protons. For a nucleon of definite mass number (A), the binding energy would be maximum at A = 2Z. If any deviation is observed from this, the binding energy term, BE_4.

$$BE_4 \propto -\frac{(A-2Z)^2}{A} = -\ a_a\ \frac{(A-2Z)^2}{A}$$

(e) Spin effect. The total value of binding energy also depends on whether the number of protons are ever or odd. It is observed that the s elf coupling of protons or neutrons gives rise to extra stability. Thus, even-even nucleus would be more stable Thus even-odd nucleus and even-odd nucleus would be more stable than odd-odd nucleus. This energy term may be put as a function of δ as follows :

$$BE_5 \propto \frac{\partial(Z, A)}{A^{3/4}} = -a_d. \frac{\partial(Z, A)}{A^{3/4}}$$

$$\text{whered } (Z, A) = \begin{cases} -1, \text{odd Z, even A (odd Z, odd N)} \\ 0, \text{odd A, odd Z, (even N, or even N, odd Z)} \\ +1, \text{even Z, even A, (even Z, even N)} \end{cases}$$

This knows as *spin effect.*

Total Binding Energy. Thus, the total binding energy may be put in the form of the following equations.

$$BE\ (MeV) = BE_1 + BE_2 + BE_3 + BE_4 + BE_5$$

$$= a_v A + (-\text{ as } A^{2/3}) + \left(-a_c \frac{Z^2}{A^{1/3}}\right) + \left(-a_c \frac{(A-2Z)^2}{A}\right) + a_d \frac{\partial(Z, A)}{A^{3/4}}$$

$$= a_v . A - a_s . A^{2/3} - a_c \frac{Z^2}{A^{3/4}} - a_a \frac{(A-2Z)^2}{A} + a_d \frac{\partial(Z, A)}{A^{3/4}}$$

where, $a_v = 14.0$; $a_s = 13.0$; $a_e = 19.3$; $a_d = 33.0$

All these constants are expressed in MeV. Therefore

$$Be\ (MeV) = 14A - 13A^{2/3} - 0.585 \frac{Z^2}{A^{1/3}} - 19.3 \frac{(A-2Z)^2}{A} + 33 \frac{\partial(Z, A)}{A^{3/4}}$$

Problem : Calculate the mass defect and binding energy per nucleon for ${}_{27}Co^{59}$. The mass of ${}_{27}Co^{59}$ as determined from mass spectrograph is 58.95182. The mass of a proton and a neutron are 1.008142 and 1.008982, respectively.

Solution:

$$\Delta M = Z.\ M_p + (A - Z)\ M_N - M(Z, A)$$

$$= 27 \times 1.008142 + 1.008982\ (59 - 27)$$

$$- 58,95182$$

$$= 0.5564381 \text{ a.m.u}$$

$$\text{Binding energy} = \frac{0.5564381 \times 931 \text{MeV}}{A}$$

$$= \frac{0.5564381 \times 931 \text{MeV}}{59}$$

$$= 8.77 \text{ MeV}$$

Semi-Empirical Mass Formula. The semi-empirical formula is represented by following equation.

$$Q = M(Z, A) - M(Z + 1, A)$$

$$= 2r\left[\pm(Z_0 - Z) - \frac{1}{2}\right]$$

$$\pm 2\delta \begin{bmatrix} +2\delta \text{ for nuclide with odd Z} \\ -2\delta \text{ fornuclide with even Z} \end{bmatrix}$$

where Q is energy of reaction of the transition $Z \longrightarrow Z \pm 1$ of isobars with mass number A ; ± d is the pairing energy ;

$$v = \frac{4a_a}{4}(1 + \frac{A^{2/3}}{4a_a / a_c}) \ ; \ Z_0 = -\frac{\beta}{2v}$$

where, $\beta = -4a_a - (M_N M_P)$, where M_N and M_P are the mass of neutron and hydrogen, respectively.

(i) The binding energies of stable nuclides.

(ii) The atomic mass M (Z, A), of nuclides with A = 4 or more; provided the values of a_v, a_a, a_s, a_c and a_d are known.

(iii) The energy of those reactions in which A undergoes change,

(iv) Energetics of the fission of heavy nuclides.

(v) Energetics of a-decay.

(vi) Energetics of transitions between isobaric nuclides.

ISOTOPES

Introduction. The term isotope was introduced by Soddy. The term isotope may be defined as follows :

"Isotopes of the element are the different atomic species of the same element whose nuclei possess the same number of protons but a different number of neutrons."

"Isotopes may also be regarded as different kinds of atoms of same element having the same atomic number, but differing in atomic weight or mass number."

Examples:

(i) There are three isotopes of oxygen found in nature. These can be represented as

$$_8O^{16},\ _8O^{17},\ _8O^{18},$$

These three isotopes of oxygen are having the same atomic number, *i.e.*, but have different mass numbers 16, 17 and 18 respectively. They possess 8, 9 and 10 neutrons in their nuclei respectively.

(ii) Chlorine (atomic number 17) is found to consist of two isotopes of mass 35 and 37. These two isotopes occur in nature in the proportion 3 : 1.

Detection of Isotopes

1. Thomson's Positive Ray Analysis. If an electric discharge is passed through a gas or vapour at low pressure, its molecules lose electrons and are converted into positive ions. These will move towards the cathode under the influence of electric field. If a cathode with the hole is used, these positive charged ions will go behind the cathode. Such a stream of positive ions is called positive rays. These rays will compose of positive charges depending upon the mass of the molecules.

Description

(i) D is the Discharge tube which is filled with the vapour of the element under examination.

(ii) C is perforated cathode made up of aluminium and is placed in the neck of the flask.

(iii) A is the anode made of aluminium and is placed in the side tube.

(iv) P and P' are two iron plates across which a strong electric field is applied. 1

(v) Plates P and P' also serve as the pole pieces of a powerful electromagnet MM. Thus, electric and magnetic fields are applied simultaneously parallel to each other but perpendicular to the path of rays.

(vi) There is a photographic plate on which positive rays are focussed.

Working. When the electric discharge is passed, positive rays are produced. The beam of the positive rays passes through the cathode and then through the plates P and P' and also through M and M. Thus, electric and magnetic fields are applied simultaneously parallel to each other but perpendicular to the path of rays.

Due to the effect of magnetic and electric fields on the positive rays, the positive ions of same mass and charge having different velocities run along a parabola line on the photographic plate. Only one parabolic line on the photographic plate is obtained if the discharge tube contains one isotope. If the discharge tube contains two or more isotopes, two or more parabolic lines will be obtained on the photographic plate. Thus, the number of parabolas on the photographic plate must be equal to the number of isotopes in the gas used in discharge tube.

If we know the parabola for a given isotope, the mass of other isotopes can be found.

Limitations of the Method. The method is not very accurate because of the following reasons :

(a) Low intensity of photographic impression. The intensity of the photographic impression is low due to the following reasons.

(i) *Because of narrow bore of the cathode tube a large number of positive rays are lost by collisions with the walls of the tube. Thus, their intensity available for the photographic plate is reduced.*

(ii) *Due to the spreading of the positive rays into a parabola, their intensity for photographic purpose if further lowered.*

(b) Blurred photographic impression. The photographic impressions are blurred with no definite edges, thus making accurate measurement difficult.

(c) Formation of secondary rays. Pressure in the apparatus, though low, yet is not negligible. As a result the positive ray particles collide with gas particles and may gain or lose electrons while passing through deflecting fields. This results in the formation of secondary rays which may be of many types. This makes the photographic impression foggy and thus making measurements more difficult.

2. Aston's Mass Spectrograph. F.A. Aston constructed an instrument called mass spectrograph with ten times the resolving power of the instrument used by Thomson.

Principle : Aston's mass spectrograph is a modification of Thomson's method. In Aston's method, the electric and magnetic fields are not applied simultaneously but one after the other in direction at right angles to each other.

The electric field produces a dispersion of the rays with respect to velocity and the magnetic field brings to focus all the particles with the same $\frac{e}{m}$ values at one point instead of spreading them in to a parabola. Parabola. In this way more sharp lines are obtained on the photographic plate.

Apparatus and Procedure

(i) The positive rays from the discharge tube first pans through two narrow slits S_1, S_2 and then through the plates P_1, P_2 to subject the positive rays to an electric field.

(ii) Rays from electric field pass between the pole pieces of a magnet M.

(iii) Now adjust the strength of electric and magnetic fields in such a way to focus the beam on a photographic plate F. The beam on the plate F is focussed in the form of a sharp line corresponding to each type of isotope present in the discharge tube.

If the discharge tube contains a large number of isotopes, a large number of lines will be obtained on the photographic plate. On developing the photographic plate a sort of mass spectrum is obtained. Due to this reason , the apparatus is called Aston's mass spectrograph.

The range of the spectrum brought on the plate can be varied by altering the strengths of the deflecting field. The mass of the particles are determined by comparison with the particle of known mass. This can be done in several ways. In the coincidence method, the intensity of the electric field is altered until the position of the known mass corresponds exactly to that originally occupied by the particles of unknown mass. Under this condition the ratio of the mass is inversely proportional to the ratio of the intensities of two fields.

(iv) *The mass of an unknown particle is determined by first putting the substances of known mass mixed with each other and their mass spectrograph obtained. Distances of lines due to known masses are measured from a particular point on the photographic plate and a calibration curve is drawn.*

Now the substances of unknown mass are mixed with another substance of known mass and spectrograph of the mixture is obtained under same conditions as before. The mass of the unknown substance is calculated from the calibration curve.

Advantage. This method is very sensitive. It is possible to determine the masses with an accuracy of the part in million.

3. Dempster's Mass Spectrometer. Dempster used a different method from Aston to produce positive ions in order to ensure that their initial energies were negligible as compared with the energy the ions acquired during their passage through the accelerating electric field. By this method it is possible to identify as well as determine the relative abundances of various isotopes of a given element.

Theory. Consider a positive ion which is accelerated from rest through a potential difference of V volts, then

$$\text{Ve} = \frac{1}{2}\text{mu}^2 \qquad \text{...(1)}$$

where e and m are charge and mass of the ion respectively and u is the velocity acquired by the ion.

If V is kept constant and all the ions have the same charge, then 1/ 2mu^2 is constant which means that all ions acquire the same energy. Such ions are called the **monoenergetic** ions.

When these monoenergetic ions are allowed to enter a uniform magnetic field and in a direction normal to the lines of magnetic force, than a force will act always at right angles to their direction of motion. Then ions will traverse a circular path of radius R given by

$$\frac{\text{mu}^2}{\text{R}} = \text{Heu}$$

where H is the uniform magnetic field.

$$\text{m} = \frac{\text{HeR}}{\text{u}} = \text{HeR}\sqrt{\frac{\text{m}}{2\text{Ve}}} \quad \text{[Use eq. (1)]}$$

$$\frac{\text{m}}{\text{e}} = \frac{\text{H}^2\text{R}^2}{2\text{V}}$$

If H and V are given, then R μ $\sqrt{\frac{\text{m}}{\text{e}}}$

i.e., ions will be sorted out in accordance with their masses.

If H and R are kept fixed, then V can be varied to make one group of ions after another to traverse a defined circular path of radius R ; the magnitude of V is required depending on the value of e/m.

Experimental details and working. P is the platinum strip on which the metallic ion to be investigated (e.g, K.Na, Mg. etc.) was spread. It was either directly heated or bombarded by electrons accelerated from thermionic filament E.

B is the large battery for production of electrostatic field. A potential difference of 500–1800 V is provided by it.

The accelerated positive ions are passed through the slit S_1 (of adjustable width) into the analysing chamber C. This chamber is made up of a heavy brass tube in which powerful magnetic field was produced between two semicircular iron plates. These plates are attracted to the surrounding electromagnets.

Only those ions which pass through screen Z and second adjustable slit S_2 were recorded by a quadrant electrometer E.

The slits provide a definite path of radius R. Then for a give value of e/m V was varied to given the appropriate value of R *i.e.*, according to equation (2), the H being kept constant.

By comparing the strengths of the ionic current produced by various isotopic particles, the relative abundances of various isotopic species in the gas or vapour under examination can be calculated.

A typical curve for the isotopes of chlorine. The heights of the peaks indicate the relative proportion of the isotopes of Cl^{35} and Cl^{37} in ordinary chlorine gas.

Limitations

(i) *We have considered that initial velocity of the ions was negligible as compared with velocity they possess in accelerating field. This is not true.*

(ii) *The resolving power of the instrument is very low of the order of* 1 *part in* 100.

4. Bainbridge Mass Spectrograph. Bainbridge in 1930 devised a mass spectrograph to modify Dempster's apparatus using photographic recording of the ions and thus much increased resolution with the great convenience of linear mass scale was obtained.

In this method the stream of ions to be subjected to the magnetic field is first made perfectly homogeneous in velocity by means of velocity filter. This produces linear mass scale.

A beam of ions of various masses is produced in a discharge tube. These ions then pass through slits S_1 and S_2 and plates P_1 and P_2, for which the displacements produced by the two fields compensate. After describing semi-circles, the ions are made to collect on a photographic plate. They will collect at different places depending upon e/m values. On the plate, traces will be obtained from which $R \propto m/e$ can be easily known. Since $m \propto R$, so it provides linear mass scale.

Theory. Let E and H' be the strength of electric and magnetic fields respectively. Force on the particle due to magnetic field = H' *eu where e and u are the charge and velocity of the particle respectively.*

Force de to electric field = eE so for all the particles coming from S_2

$$H' \, eu = eE$$

$$u = \frac{E}{H'}, \text{ where E and H' are constants.}$$

A monovelocity beam of ions enters the analysing chamber, where a magnetic field H acts perpendicular to the trajectories of the ions. In this analysing chamber, an ion of mass m traverses a semi-circular path of radius R as given by

$$\frac{mu^2}{R} = H\,eu$$

$$R = \frac{mu}{He} = \frac{mE}{HH'e}$$

Because u is constant irrespective of ion therefore

$$R = km$$

where k is constant. H is kept constant, for the ions of the same charge e.

From the above relation

$$\frac{e}{m} = \frac{E}{HH'R}$$

$$\frac{e}{m} \mu \frac{1}{R} \text{ since E, H, H' are constants.}$$

Limitation

For evaluating the relative abundances of isotopes this method appears to be less suitable than the Aston's because of velocity selection.

5. E. Nier's Mass Spectrograph (1935). This involves the principle of double focussing, *i.e.*, both direction and velocity focussing.

This is simply an improvement over Dempster's mass spectrograph and is used for the measurements of the relative abundances of isotopes; important improvements are given below :

Apparatus and procedure. A systematic diagram of Nier's mass spectrograph.

(1) *The positive ions are produced by he ionisation of vapours of the element under electron bombardment in a main tube.*

(ii) The positive ions are accelerated by an electric field of known accelerating potential, V, applied between the two charged plates P_1 and P_2 and are allowed to pass though the slit S_1.

(iii) After slit S1 rays are allowed to flow down a copper tube and then deflected through an angle of 60° by applying magnetic field M.

(iv) From the magnetic field, positive rays are allowed to pass through another slit S_2 after which the beam comes to a focus on the collector E which is connected to a suitable amplifying device to measure the ionic current.

(v) The apparatus is designed in such a way that the ions with different e/m values can be brought to focus at the collector E by changing the accelerating potential V. Since the latter is proportional to e/m their abundances can be determined easily by the magnitude of current.

Advantages

(i) On account of small ionic current, isotopes of even very low abundance can be determined. It can measure the abundance as low as 1 in 10,000 parts.

(ii) The Nier ion source produces positive ions under precisely controlled conditions which justifies the accuracy of the result.

SEPARATION OF ISOTOPES

Separation Factor. *The extent to which a mixture of two isotopes may be separated in one stage of a particular process is* called the separation factor. It is the ratio of the relative concentration of the desired isotope after processing to its relative concentration before processing. The factor, therefore, is a quantitative value for a one stage process indicating the efficiency of the separation.

In mathematical notation, if n_1 and n_2 are the numbers (or moles) of light and heavy species in a mixture before processing and n_3 and n_4 are the corresponding quantities after processing, in a one-stage process., the separation factor s may be indicated as

$$s = \frac{n_3/n_4}{n_1/n_2}$$

Usually this factor is only slightly greater than unity, and consequently a useful separation process is made up of many single separation states. In such a case the separation factor is *s* for each stage, and if the entire process is of *x* states, then the over-all separation factor S is given by

$$S = s^x$$

The efficiency of the process is evaluated by the magnitude of its overall separation factor.

Different methods for the separation of isotopes are :

1. Gaseous Diffusion Method. According to Graham's law of gaseous diffusion, the rate of diffusion of a gas is inversely proportional to the square root of its density. It means that if a mixture of gaseous isotopes is allowed to diffuse through a porous vessel, the lighter isotope will diffuse faster than the heavier one. The greater the ratio of masses of two constituent isotopes, the more efficient will be the separation. Consequently this method is more efficient for the separation of lighter isotopic elements.

Aston first attempted to separate neon isotopes by making use of this method and he succeeded in concentrating neon to some extent. However this method is very difficult.

The efficiency of the diffusion process was greatly increased by O. Hertz (1932) by utilising *Cascade principle* in which partially enriched output of one stage is automatically made the input of the second stage and the output of the second stage for the next and so on.

Hertz (1932) by using 48 diffusion pumps in series was able to get 99.9% heavy hydrogen This method has been used for the separation of isotopes of carbon, nitrogen, oxygen and argon.

By using uranium hexafluoride (UF_6) as the diffusing gas and employing a few thousand stages, the enrichment of lighter isotope of U^{235} has been done.

2. Thermal Diffusion Method. If a gaseous mixture of light and heavy isotopes is enclosed in a space between two vertical plates maintained at different temperatures, the two processes will take place simultaneously. These are :

(i) Due to the thermal diffusion, the molecules containing heavier isotope will move towards the cooler surface and lighter towards the hot surface.

(ii) At the same time convection currents are set up which result in the movement of lighter isotope towards the top and the heavier towards the bottom.

The net effect of (i) and (ii) is to transfer the lighter isotope towards the top and heavier towards the bottom.

Apparatus and working. The apparatus consists of vertical tube about three meters in length and one cm in diameter. It is heated by enclosing an electric heating coil within it. From outside it is kept cold by circulating cold water. The temperature difference between the inner and outer surface of the tube is maintained at about 500°C.

On passing the gas of the isotopes to be separated through the tube, the lighter isotope concentrates around the hot wire while the heavier isotope concentrates around the cold surface of the tube.

The convection currents, set up in the tube, carry the lighter isotope upwards, while the heavier one is carried downwards. The two fractions are collected in the collecting flasks placed at the two ends of the tube. Clausius and Dickel applied this principle for the separation of isotopes of chlorine, carbon, argon and neon.

For enrichment of isotopes on a laboratory stage, this method has advantages over other methods because of its simplicity and efficiency.

Using several units in series. HCl was separated almost completely into H^{35} Cl and H^{37}Cl, the latter being obtained in 99.% purity.

3. Electromagnetic Method (Mass spectrograph method). When positive rays obtained from an isotopic mixture of an element is subjected to simultaneous electric and magnetic fields, the various isotopes separate according to their mass numbers. By placing collecting chambers at calculated positions, it is possible to obtain different isotopes in pure form.

Lawrence (1946) used this method on a large scale to get U^{235} isotopes for its use in atomic bomb.

The electromagnetic method has a high separation factor but the operation is not simple. In developing this process, three limitations were encountered. These are :

(i) It is difficult to produce a large quantity of gaseous ions.

(ii) In producing the narrow beam by means of a slit system, only a small fraction of the total ions can be utilized.

(iii) As the quantity of material in the beam is increased, there is a corresponding increase in space - charge effects which interfere with the separating action.

However, the first attempt to improve the design of this method to minimise the foregoing limitations was performed by E.O. Lawrence and his coworkers. They built their apparatus around an electromagnet removed from a cyclotron, and was called a calutron. It was soon realised that the efficiency of the method could greatly be increased if an enriched material were fed into the units.

4. Centrifugal Method. When an isotopic mixture is put into an ultracentrifuge which is rotated at very high speed, the heavier isotope will concentrate near the periphery while the lighter isotopic fraction near the axis of rotation.

The extent of separation is theoretically dependent upon the **mass difference** between the isotopes, and not on their ratio. This method has an attractive feature that the separation is fully as effective for the isotopes of heavier elements as for the lighter ones. This method was suggested in 1919 by Lindemann and Aston. It was later developed in 1922 by Mulliken.

In 1939 J.W. Beams utilised this method for separating isotopes of chlorine by using high - velocity gas drive centrifuge. In an applied centrifugal field, the theoretical separation factor S (*i.e.*, ratio of the isotopes at the centre over the corresponding ration at the periphery) is given by

$$S = e^{V_2 (M_1 - M_2)/2Rt}$$

where V = Peripheral velocity,

M_2 and M_1 = molecular weights of the materials containing the isotopes under investigation.

R = Gas constant, and

T = Absolute temperature

Appreciable concentrations of uranium isotopes have been done by using tall cylindrical centrifuges.

5. Ion Migration Method. This method is based upon the simple principle that ions of different masses should migrate with different velocities (Lindemann's Method). This method was used for increasing the concentration of K^{39} in the mixture of K^{39} and K^{40}. Electrolytic ion migration in fused melts has been utilised to increase the concentration of the heavier isotopes of silver, lithium and potassium. This method could not do the complete separation but to only the concentration.

6. Distillation Method. When a liquid consisting of a mixture of isotopes is evaporated at low pressure and the evaporated molecules are immediately condensed, the condensate will be richer in the lighter isotopes while the residual liquid in the distillation flask will become richer in the heavier component.

Bramsted and Hevsey (1921) utilised this principle for the separation of isotopes of mercury, neon, etc. Distillation method is used to separate nuclides from an isotopic mixture if they are available as liquid compounds with sufficiently different vapour pressures. The separation factor is equal to the ratio of the vapour pressure at the operating temperature of the two components. It is only slightly greater than unity.

To increase the efficiency of the process, counter current flow is usually employed in the distillation columns. Vapour travels upwards and is arranged to intermingle closely with the downward moving liquid stream.

Urey and his associates used distillation to establish the existence of deuterium.

The distillation method has been used to some extent to produce deuterium but does not compete with the electrolytic method or the chemical exchange method. It has been used to produce water enriched with oxygen-18.

7. Fractional Electrolysis. When an isotopic mixture is electrolysed, the ions of heavier isotopes are comparatively sluggish and liberated at the cathode later than the lighter isotopes. This method was applied to isolate heavy hydrogen (Deuterium $_1D^2$) from water by Urey (1933). He electrolysed 0.5 MNaOH solution using large currents (30 – 300 amperes/

hour) and nickel electrodes till the volume was reduced to 1/10th of the original. This process was repeated several times till water containing 99% D_2O was obtained.

The current consumption is large in this process. Thirty to forty thousand ampere-hours or electricity are consumed in producing of heavy water.

The electrolytic method is only satisfactory in practice for separating the isotopes of hydrogen of mass numbers 1 and 2.

Lewis and Macdonald (1936) separated lithium isotopes by using the difference in electrode potentials of the two isotopes.

8. Chemical Exchange Method. It has been noted that in many exchange reactions, the rates of reactions of different isotopes are different. This was treated theoretically by Urey and Oreiff. When a substance in gaseous states is in equilibrium with its liquid state, the isotopic composition of two phases is different.

$$2H_2O^{18} + O_2^{16} \rightleftharpoons 2H_2O^{16} + O_2^{18}.$$

It means that of ordinary oxygen (mixture of O^{16} and O^{18}) is kept for sufficient time with water (containing H_2O^{16} and H_2O^{18}) the heavier isotope O^{18} will be more in the gaseous phase than in water phase. Thus, oxygen gas richer in O^{18} can be obtained.

Some other examples of isotopic equilibria in the gaseous and liquid states are :

$$O_2^{16}\ (g) + 2H_2O^{18} \rightleftharpoons SO_2^{18}\ (g) + 2H_2O^{16}\ (l)$$

$$N^{15}O + N^{14}O_2 \rightleftharpoons N^{14}O + N^{15}O_2$$

$$Cl_2^{35} + 2HCl^{37} \rightleftharpoons Cl_2^{37} + 2HCl^{35}$$

The possibility of separation of oxygen isotopes by ion hydration equilibria was first demonstrated experimentally by Feder and Taube (1952) :

$$A(H_2O)_n + H_2O^* \rightleftharpoons A(H_2O)_{n-1}\ (H_2O^*) + H_2O$$

For the successful operation of this process the following conditions are to be kept in mind.

(i) Rapid establishment of the equilibrium,

(ii) Intimate contact between the phases and

(iii) Ready conversion of gaseous component containing the desired element into the liquid component and vice versa, to render the process continuous.

This method is particularly advantageous in that it can be utilised without difficulty to increase the separation factors remarkably. But in this method, large scale separation is made possible.

9. Ion-exchange and Chromatographic Methods. By using the following exchangers of the type

$$\text{Na} - \text{Zeolite} + \underset{(\text{Soln.})}{\text{Li}^+} \rightleftharpoons \underset{(\text{Soln.})}{\text{Na}^+} + \text{Li} - \text{Zeolite}$$

there occurs preferential intake of ^{6}Li into the zeolite phase leading to a concentration of ^{6}Li in the adsorbed Li^+ ; the LiCl solution was used. By using NH_4Cl solution in a similar manner, ^{15}N was concentrated on the zeolite phase.

Enrichment of potassium and of nitrogen isotopes by chromatographic adsorption on substances like $BaSO_4$, silica gel, etc. has also been reported: the heavier isotopes in these cases concentrate in the adsorbed fraction.

The separation of neon isotopes by adsorption charcoal (^{22}Ne enriching in the gas phase) has been reported. Also, the complete separation of hydrogen isotopes on palladium black mixed with purified asbestos (H^2 enriching in the gas phase) has also been reported.

10. Photochemical method. This method was used by Kuhn and Martin to separate chlorine isotopes by using $COCl_2$. Light of wavelength 2816Å and 1790Å is absorbed by $COCl_2$. This light is obtained form aluminium spectrum by employing suitable filters. This light illuminates $COCl_2$ at 1000 mm pressure of mercury in presence of a trace of iodine when the following reactions take place.

$$COCl_2 \rightarrow CO + Cl_2$$

$$Cl_2 + HgI_2 \rightarrow HgCI2 + I_2$$

In 2 months, 2g of $HgCl_2$ enriched in ^{35}Cl is obtained which possesses chlorine of atomic weight 35.430 (as opposed to 35.455 for ordinary chlorine).

COMPOSITION OF NUCLEUS

(1) Electron-proton concept. Prior to 1932, the date of discovery of the neutron, it was generally believed that nuclei were composed of protons and electrons. In this model the nucleus was considered to contain sufficient proton's to account for its atomic mass and sufficient electrons to reduce the net positive charge to a value equivalent to the number of planetary electrons. Thus, if an element of atomic weight A has an atomic number Z, its nucleus would contain A protons and A–Z electrons.

Evidence : One strong argument in favour of the above model was the emission of beta particles [electrons] by certain radioactive elements which were supposed to be coming from the nucleus.

Success of the Theory. This theory could explain the emission of α-and (β-particles. Presence of electrons in the nucleus showed that under the appropriate conditions one of them might be ejected in the form of β-particle and also two electrons could combine with four protons to form α-particle before emitting. This theory led to a number of contradictions with experiments.

Objections : This theory could not explain the following points :

(i) The size of an electron is approximately same as that of average nucleus, therefore it was impossible to pack so many large particles into a single body of the size of one electron.

(ii) It could not explain the angular momentum of the nuclei.

(iii) This theory could not explain the dual β-decay (e^- and e^+) exhibited in many nuclides.

(iv) This theory could not explain the Fermi's interpretation of β-decay in terms of the emission of an electron of an electron-neutrino pair.

(v) Several theoretical considerations reveal that a bound fundamental panicle *i.e.*, electrons and protons) cannot localised in the region smaller than its compton wave length h/mc.

Compton wavelength of the electron

$$= \frac{6.6 \times 10^{34}}{9 \times 10^{-31} \times 3 \times 10^{8}} = 250 \times 10^{-14} = 2.5 \times 10^{-12}$$

This rules out the possibility of keeping the electrons inside the nucleus.

(2) Proton-neutron concept. The discovery of the neutron by Chad wick in 1932 led to the rationalisation of a new nuclear model to accommodate the neutral particle, *i.e.*, the neutron. In this new model nuclei are composed of protons and neutrons. There must be enough protons to account for its charge. The number of neutrons is equal to the difference of atomic mass and atomic number. The sum of the protons and neutrons is considered as giving approximately the mass of the nucleus. The number of protons and neutrons in the nuclei of the atoms of atomic weights are nearly equal, but the relative number of neutrons becomes increasingly great for elements of higher atomic weights.

The atomic weight of uranium, for example, is 238, and its atomic number is 92 indicating that the nucleus of uranium atom is composed of 92 protons and 146 neutrons. On the other hand, the atomic weight of oxygen is 16, and its atomic number is 8, therefore the protons and neutrons in the nucleus must be equal in number.

The following points support proton – neutron theory.

(i) Spin considerations. Both protons and neutrons have the same spin quantum number 1/2. So according as A is odd or even the resultant spin of A nucleons will be an integral of half integral multiple of k. This again agrees with the experimental observations. Since $m_n = m_p$, the value of magnetic moment of the neutron is approximately equal to that of protons. The magnetic moment of all nuclei as measured are consistent with these values

(ii) Nuclear size. The total energy of the proton is approximately 940 Mev from momentum space uncertainty considerations. Since the rest energy of proton is 930 KeV, the kinetic energy of neutron or proton in the nucleus is of the order of few MeV arid hence a free proton or neutron may reside in the nucleus.

(iii) *It explains the dual* β*-decay.* *The electron does not exist in the nucleus but it is formed at the time of emission as indicated by the following equation :*

$$n \rightarrow p + e^- + v$$

β^+ decay is due to the following reactions :

$$p \rightarrow n + e^+ + v$$

(iv) *Compton wavelength of neutron* (λ) *is given by the following relation :*

$$\lambda = \frac{h}{mc} = \frac{6.6 \times 10^{-34}}{1.67 \times 10^{-27} \times 3 \times 10^8} = 0.14 \text{ (Nuclear diameter)}$$

Thus, the protons and neutrons could be accommodated into nuclear volume.

(v) *Both the protons and neutron have fame spin quantum number, 1/2. Therefore, according to the quantum theory, the resultant spin of A nucleus will be an integral or half - integral multiple of* $\frac{h}{2\pi}$ *according as A is even or odd. This is in agreement with all the experimental observations.*

(3) **Neutron–positron Concept.** This was given by Jean Perrin. According to this concept, the atomic nuclei are built up of neutrons and positrons only.

This concept goes deeper into the structure of atomic nuclei. However, it suffers from the same difficulties with electron – proton concept. Comes across.

$$p \rightarrow n + e^-$$

(4) Antiproton–neutron concept. This is a more recent concept which has been suggested by Gamow, Klin, etc. According to this concept, the atomic nuclei consist of antiprotons and neutrons.

It is interesting to note that the relation between the negative protons and ordinary protons is analogous to that between positrons and electrons in Dirac's theory. If a neutron is converted into anti-proton, there occurs the emission of a positron.

$$n \rightarrow p^- + e^+$$

An antiproton may be conceived to be formed by a neutron and electron.

$$p^+ \rightarrow n + e^-$$

Similar to positrons, antiprotons cannot exist free for long within the ordinary material as they will be immediately attracted arid absorbed by nearest +vely charged nucleus: Energy of about 4000 MeV is essential to create a pair of proton and antiproton.

Note: The neutron–proton model is still the accepted one.

ELEMENTARY IDEAS ABOUT NUCLEAR FORCES

There is a large amount of experimental information available about the fact that the atomic nuclei are built of protons and neutrons, but as yet there is no satisfactory explanation about the nature of forces that bind the nucleons together and keep the nuclei stable.

The nature of forces cannot be ordinarily electro-static because the particles involved have either positive charge or no charge. Also the forces in question cannot be gravitational because these forces are even weaker than electrostatic forces.

Properties

1. **Saturation Property.** The nuclei have same density and the binding energy per nucleon for nuclei with A > 40 is constant. Nucleons attract each other strongly only if they are in die same orbital state. As a result, each nucleon interacts with only a limited number of nucleons nearest to it. This is referred to as the saturation property of nuclear forces.
2. **Charge independence.** The nuclear forces acting between two protons or between two neutrons or between a proton and a neutron are same. It follows that the nuclear forces are non-electric in nature.
3. **Nuclear forces are short range forces.** Nuclear forces are appreciable only when the distance between the nucleons is of the order of 10^{-15} m or less. These distances are called the action radii or range of the nuclear forces. The interaction between nucleons is accomplished by the exchange of pi-mesons.

Let m be the rest mass of the pi-meson. $\Delta E = mc^2$ is the rest mass energy of the π- meson. According to Heisenberg's uncertainty principle the time required for nucleons to exchange π- mesons cannot exceed At, for which.

$$\Delta E\ \Delta t \geq k$$

The distance π - meson can travel away from a nucleon in the nucleus during the time Δt, even at a velocity ≈ c *is* $R_0 \approx k/m_e = 1.2 \times 10^{-15}$m. This coincides with the value of the nuclear radius and is of the order of magnitude of the nuclear range.

4. **Nuclear forces are not central forces.** In particular, they depend on the orientation of the spin.

From the above discussion it is evident that there must be entirely new type of mechanism involved to account for the strong attraction between nucleons when they are very close to each other. Some theories describing me nature of these forces have been discussed below :

(i) Exchange Theory. Heisenberg and Mayorana proposed the first theory about nuclear forces by suitably combining the liquid drop and shell models and bringing in an idea of what is called an exchange force between the nucleons. They proposed that : " *When a neutron interacts with a proton, a single electric charge jumps from one nucleon to the other so that in the jump the original proton changes to a neutron and the neutron into a proton*".

Fermi based his theory of β- decay on this exchange idea. According to this theory, a neutron can change to a proton by emitting an electron and neutrino.

$$\text{Neutron} \rightarrow \text{proton} + \text{electron} + \text{neutrino}$$

This theory of Fermi was successful in explaining continuous β-spectra. From this explanation, Heisenberg got the idea that *the nuclear forces are exchange forces in which electrons or positrons and neutrinos are exchanged between the nuclear particles.*

Mavaorana modified the above theory of exchange forces by suggesting that it is not only the electric charge that exchanges, but also the spin of the particles is exchanged.

It is to be remembered that the nuclear forces are effective at short distances and their magnitude changes with the distance. The effective forces Are maximum at a distance of 8.0×10^{-14} cm and convert into that of repulsion when the distance is 5.0×10^{-14}. The attractive force is zero at a distance of about 4×10^{-13}cm. ;

Objective. From the p-decay studies, the magnitude of the nuclear forces was calculated and it was found that the exchange forces that result from the electron - neutron exchange are weaker by a factor of about 10^{-14} as those required theoretically. Thus, although theory was simple, it failed to work.

(ii) Yukawa Theory. Yukawa in 1935 predicted the existence of a new particle and modified the exchange theory by proposing that the exchange particles are not the electron and neutrino but the new particle *meson* which has rest mass between electron and proton.

According to Yukawa, when a proton and neutron interact, the proton may emit a positive meson which is absorbed by the neutron, therefore, in the exchange of the positive meson, the proton becomes a neutron, and the neutron becomes a proton. In like manner a neutron may interact with a proton by emitting a negative meson, and in the process the neutron becomes a proton and the proton becomes a neutron. These two interactions may be represented as

$$p = \pi^{+} + n$$

$$n = \pi^{-} + p$$

Mesons of widely different masses and life times were discovered and there arose a question as to which of those was the exchange particle. Yukawa himself took the exchange particle to be μ meson which has mass 200 times the rest mass of electron. Latest researches have shown that it is the π meson which is the exchange particle.

Success : Yukawa's theory provides an explanation for the binding forces between protons and neutrons and *vice versa.*

Objection : Yukawa's theory does not account for forces between like particles. A proton cannot absorb a positive meson to acquire a second positive charge, consequently a charged meson cannot explain the proton - proton bond. Also it is unlikely that a neutron can absorb a negative meson in the formation of a neutron - neutron bond.

Yukawa potential and nuclear force. The important features of the nuclear force is its range. That is the nuclear force decreases extremely rapidly when the interacting nucleons are separated, beyond 1 Fermi. Experiments show that there is a critical length beyond which the interaction does not extend. In this aspect, it differs fundamentally from a $1/r^2$ force such as the electromagnetic force. We must thus expect, if there is a nuclear potential, it will contain a parameter with the dimension of the length. Actually such a potential was first proposed by Yukawa and is called Yukawa potential. It is of the form.

$$V(r) = -g\frac{e^{-kr}}{r}. \qquad \ldots (1)$$

Where g is a constant of interaction and k is the reciprocal of the length which can be assumed to represent the range of the nuclear force.

Using uncertainly principle, Yukawa showed that the mass of a meson is 200 times greater than the mass of an electron.

Actual mass of the charged pion is 273 m_e, while the mass of neutral pion is 264 m_e.

Yukawa Theory. The relationship between the total energy *E,* the momentum *P* and the rest mass m_e of the particle which is relative is

$$E^2 = P^2c^2 + m_0^2c^4 \qquad \ldots(2)$$

Where c is the velocity of light in free space. If *E* and *P* are replaced by its quantum mechanical operators.

$$\left.\begin{array}{l}\dfrac{ih}{2\pi}\partial/\partial t \\ \dfrac{-ih}{2\pi}\partial/\partial r\end{array}\right\} \quad \text{...(3)}$$

It terms of operators the above relation is written as

$$\frac{-h^2}{4\pi^2}\frac{\partial^2}{\partial r^2} = \frac{-h^2}{4\pi^2}e^2\frac{\partial^2}{\partial r^2} + m_0{}^2 c^4$$

Using the Laplacian operation,

$$\nabla^2 = \frac{\partial_2}{\partial r^2} = \frac{\partial^2}{\partial x^2} + \frac{\partial^2}{\partial y^2} + \frac{\partial^2}{\partial z^2}$$

The above relation can be written as

$$\nabla^2 = \frac{1}{c^2}\frac{\partial^2}{\partial t^2} - \frac{4\pi^2 m_0 c^4}{h^2} = 0$$

Let us no introduce a potential function

$$\phi = (r, t) = \phi(r)\,\phi(t)$$

$$\nabla^2\phi = \frac{1}{c^2}\frac{\partial^2}{\partial t^2} - \frac{4\pi^2 m_0{}^2 c^4}{h^2}\phi = 0 \quad \text{...(4)}$$

This is known as Klein–Gordon equation. If $m_0 = 0$, the rest mass of the mass of the particle exchanged is zero.

∴ The exchanged particle is a photon. As photons are exchanged when two particles interact with electromagnetic interaction the mesons are exchanged in strong interaction between nucleons. The time independent part of this eqn. can be obtained by setting $\frac{\partial\phi}{\partial t} = 0$, hence

$$\nabla^2 f - \frac{4\pi^2 m_0{}^2 c^4}{h^2} f = 0$$

$$\frac{4\pi^2 m_0{}^2 c^4}{h^2}$$

$$(\nabla^2 - K^2)\phi = 0 \quad \text{...(5)}$$

Where $K = \frac{4\pi^2 m_0{}^2 c^4}{h^2} = \frac{1}{r_0}$ a characteristic constant of the Yukawa potential.

This equation is analogous to Laplaces's equation ($\nabla^2\phi = 0$) valid in the case of electromagnetic field in the absence of electric charge. In the presence of charges the equation will be analogous to Poisson's ($\nabla^2\phi$ ρ/ε_0).

Hence for meson field in the presence of nucleon, we have

$$(\nabla^2 - K^2)\ \phi = g$$

Where 'g' is the nucleon charge which measures the strength of the interaction between the nuelcon and the meson field.

The solution of the equation becomes $\phi = \frac{-ge^{-kr}}{r}$ ...(6)

This potential function is called Yukawa potential. The variation of this potential with the distance between two nucleons. It clearly indicates the short range nature of nuclear force. The resultant spin will be an integral spin. The observed experimental values of nuclear spin are in good agreement with this prediction.

(iii) Modification of Yukawa's Theory. To take care of binding forces between like particles as described above, N. Kemmer the English physicist reasoned that there must be a neutral meson The neutral meson has since been discovered. The interactions between like particles may be indicated as

$$P = \pi^0 + p$$

$$n = \pi^0 + n$$

The neutral meson might also serve to carry forces between unlike particles as well as between like particles.

Note. There is no simple mathematical way to demonstrate, how the exchange of particles between two bodies can lead to attractive forces.

(iv) Nuclear Fluid Theory. According to this theory the nucleons are present in the nucleus (of the order 10^{-12} to 10^{-13} cm) in the form of a nuclear fluid which has exceptionally high density *i.e.*, about 130 million tons/ml (about 100 trillion times that of water). Due to this high density, the surface tension of the nuclear fluid is very high. This indicates that the space and distance between particles in the nucleus are very small. All nuclei have a uniform density, *i.e.*, the density of nuclei does not vary from atom to atom but is a constant quantity. It is because of this high density, *i.e.*, surface tension, that the nucleons are bound together despite the presence of repulsive forces.

MODELS OF THE NUCLEUS

There are many isolated facts which will require explanation when we adopt any nuclear model. Some of these facts are ?is follows :

(i) Why are nuclei emitting–particles when these only contain protons and neutrons;

(ii) Why is the binding energy per nucleon constant ?

(iii) why are the 4n nuclei particularly stable?

(iv) What is the explanation for the existence of excited states of nuclei

(v) What is the explanation for the Geiger – Nutall's ;rule ?

(vi) How are the special properties' (viz., stability, spin, magnetic moment, etc.) of nuclei explained ?

It is essential to have a model of the nucleus if many experimental observations on nuclear radiations and interactions are to be explained on a fundamental basis. An entirely satisfactory nuclear model has lot yet been put forwarded. However, various models Have been proposed. Of these, only four need to be considered; they are :

(a) Fermi Gas Model. According to this model, the nucleus is considered to be composed of a degenerate Fermi gas of neutrons and protons. The Fermi gas has been assumed to be degenerate as all he particles are present into the lowest possible states in a manner which is consistent with the requirements of the Pauli's principle. For each type of particle, the gas may be characterised by the K.E. of the highest filled state, the Fermi energy. The nucleons are moving freely with in a spherical potential will of the proper diameter so adjusted that the Fermi energy raises the highest lying, nucleons upto the observed binding energy.

The Fermi gas model is not useful for the prediction of the detailed properties of low lying states if nuclei observed in the radioactive decay processes.

(b) α-Particle Model. The model is based on the fact that α-particles are ejected by nuclei in disintegration, both in natural and artificial radioactivity. According to this model, a- particles are present within the nucleus except, of course, in the case of simplest atoms Of hydrogen and deuterium.

This model has been successful in explaining some, of the, nuclear phenomena exhibited by the light elements and in explaining the α-particle emission by. Heavy elements.

Objections. It was soon realised that α- particles cannot maintain their identity for a long time inside condensed nuclear matter and will dissolve into more elementary particles.

(c) Liquid Drop Model. In the liquid' drop model proposed by Bohr in 1937 it is suggested that the nucleus is analogous in certain respects with a small electrically charged drop of liquid. The following analogies hold between i a small drop of liquid a nucleus.

(i) The drop is spherical because of the symmetrical surface tension forces which act towards the centre. The nucleus is assumed to be spherical.

(ii) The density a spherical drop is independent of its volume. This is also the case for a nucleus. However, there is a disparity. Whereas the density of nuclear matter is independent of the type of nucleus, the density of a liquid depends on the type of liquid A given liquid, say water, must therefore be considered in the analogy.

(iii) The molecules in a liquid drop model interact over short ranges compared with the diameter of the drop, Like the nucleons in a nucleus, the molecules in a liquid drop interact only with their immediate neighbours.

(iv) The surface tension forces acting at the surface of a drop may be compared with the potential barrier effect at the surface of a nucleus.

(V) The molecules in the drop move short distances with thermal velocities. If the 'thermal agitation is increased by raising' the temperature, evaporation of molecules takes place. The nucleons in a nucleus also have kinetic energy. If energy is given to the nucleus by a bombarding particle, a compound nucleus is formed which emits nucleons almost immediately.

(vi) If a drop is made to oscillate, it tends to separate into two pars of equal size. The capture of neutrons by the nuclei of certain heavy elements leads to nuclear fission in which the nucleos breaks up into the two for fragments of roughly equal size.

Inspite of these similarities, there are some dissimilarities between a nucleus and a liquid drop.

(i) Molecules of a drop attract one another at distances which are large than the dimensions of the electron shells and repel strongly when the distance is smaller than the size of electron orbitals.

(ii) The average K.E. of the molecules in the liquid is of the order of 0.1 eV while the corresponding de Broglie's wavelength is 5×10^{-11} m which is very much smaller than the intermolecular distances. On the other hand, the average kinetic energy of nucleons is of the order of 10 meV while the corresponding de Broglie's wavelength is 6×10^{-15} m which is of the order of intermolecular distances. Thus, the motion of the molecule in the liquid is of classical mechanics while in nuclei in the motion of the nucleons is of quantum character.

In like manner one can picture the building of the structure of the nucleus as the gradual filling up of single particle orbits by neutrons and protons. The orbits can he described by the same letters as those used to designate the quantized orbital angular momentum of electrons; s = 0, p = 1, d = 2, f = 3, g = 4, h = 5, i = 6. Since the neutrons arid protons obey the ^Pauli exclusion principle, the s level has room for just 2 protons and 2 neutrons, the p level has room for 6 protons and 6 neutrons. One of the proposed quantum-configuration is given in Table 2.1. In this table the .ordinary numbers refer to shells, the lower case letters designate the orbitals , and the superscript numbers indicate the number of nucleons in each orbital.

Table 2.1 : Closed shell structures in the nucleus.

Shell	*Configuration*
2	$1s^2$
8	$1s^2$, $2p^6$
20	$1s^2$ $2p^6$ $2s^2$ $3d^{10}$
50	$1s^2$ $2p^6$ d^{10} $4f^{14}$ $5g^{18}$

The first important consequence of the nuclear droplet theory is that the volumes of different atomic nuclei must be proportional to their mass since the density of the fluid always remains the same, regardless of the size of the droplets which it forms. The conclusion is completely confirmed by direct measurement of nuclear radii which show that throughout the entire sequence of elements, the radii of atomic nuclei vary as the cube of their mass.

$$M = \frac{4\pi}{3} r^3 p$$

Where ρ is the density of nuclear fluid.

The main drawbacks of the liquid drop model are that the density and surface tension of the nuclear fluid have fantastic values. Let us calculate the density of nuclear fluid from the data for oxygen nucleus.

$$\text{Nuclear density} = \frac{\text{Mass of oxygen}}{\frac{4\pi}{3}(\text{radius of oxygen})^3}$$

$$= \frac{2.66 \times 10^{-24}\,\text{gm}}{\frac{4\pi}{3}(3 \times 10^{-13})^3\,\text{cm}^3}$$

$$\text{Nuclear density} = \frac{2.66 \times 10^{24}}{1.13 \times 10^{-27}} = 2.4 \times 10^{14}\ \text{gm/cm}^3$$

This is very high density indeed. If the nuclear fluid which is depressed through space in the form of minute droplets surrounded by rarefied electronic envelopes, could be collected to form a continuous material, one cubic centimeter of it would have two hundred and forty million ions.

Along with its almost unbelievable high density, nuclear fluid possesses a corresponding high surface tension. The surface tension of nuclear fluid is found to be

93, 000, 000, 000, 000, 000, 000, 000, dynes/cm.

The liquid drop model is found very useful in the study of nuclear fission and fusion.

The Magic Number. It has become apparent in recent years that many nuclear properties vary periodically in a sense similar to that of the periodic system of the elements.

Most of the properties show marked discontinuites near certain even values of the proton or neutron number. It has been observed that atoms with an even number of nucleons in their nuclei are more plentiful than those with an odd number. This would indicate that a nucleus made up of even nucleons is more stable than one having odd nucleons.

The experimental facts have revealed that especially stable nuclei result when either the number of neutrons or protons is equal to one of the number 2, 8, 20, 50, 82, 126. These numbers are commonly referred to as **magic numbers.** These .magic numbers can be arranged into two series.

A 2, 8, 20 (40).

B 2, (60, (14), (28), 50, 82, 126

The first series, A, can be represented by the formula (n + 1) (n + 2) /3 where n is an integer; the second series *B by n* (n^2 + 5)/5. The series A is used for light nuclei, the series B for heavy nuclei. The figures in brackets are so called the semi-magic numbers ; they correspond to less stable nuclei.

In the series A, nuclides containing 2, 8 or 20 protons or neutrons will consequently be more stable than their immediate neighbours. In series B, heavy nuclides containing 2, 50 or 82 protons or neutrons or 126 neutrons will be extra stable.

Following informations led to the prediction of the existence of energy levels or shells in the nucleus :

(i) The atoms having magic number nucleons have been found to be stable and abundant than those atoms with nucleons above and below magic numbers.

(ii) The binding energy of atoms having nucleons corresponding to magic numbers has been found to be greater than their neighbours.

(iii) α- emitting radioactive atoms tend to attain the stable configuration associated with magic number totals.

(iv) The atoms having nucleons just above the magic numbers are less table Y extremely short –lived), occur only as intermediate products in processes which involve artificial radioactivity and are neutron emitters; after emitting neutrons they tend to attain magic number totals.

Like the building of electron orbital in the atom, the building of structure of nucleus can be conceived and have been found to follow the same rules. In this respect a nucleon (a proton and a neutron) is similar to electron and obeys the Exclusion Principle. Thus, an *s* level may have two nucleons (*i.e.*, 2 protons and 2 neutrons), *p* - level may have a room for six nucleons, *d* foı ten and of f for fourteen nucleons. Thus, the arrangement of nucleons in the nuclear shell with magic number totals can be given as follows :

Table 1

Shell	*Configuration*
2	$1s^2$
8	$1s^2\ 2p^6$
20	$1s^2\ 2p^6\ 2s^2\ 3d^{10}$
50	$1s^2\ 2p^6\ 3d^{10}\ 4f^{14}\ 5g^{18}$
82	$1s^2\ 2p^6\ 3d^{10}\ 4f^{14}\ 5g^{18}\ 6h^{22}\ 4d^{10}$
126	$1s^2\ 2p^6\ 3d^{10}\ 4f^{14}\ 5g^{18}\ 6h^{22}\ 7i^{26}\ 5f^{14}\ 3p^6\ 4p^6\ 2s^2$

Evidences for the existence of the Magic Number

(i) Nuclei containing a magic number of neutrons exhibit low cross sections for the capture of neutrons of moderate energy This is explained on the basis that such nuclei have closed neutron shells and cannot therefore readily hold an extra neutron.

(ii) Evidence for the magic numbers is provided by measurement of perturbation in the hyperfine structure of the optical spectrum lines of the elements.

(iii) Additional evidence for the existence of the magic numbers is provided by a study of nuclear spin, and the phenomenon of nuclear fission.

Successes of the Liquid Drop Model. These are :

(i) This forms the basis of Bohr's theory of the compound nucleus formation in nuclear reactions.

(ii) This model gives atomic masses and binding energies accurately.

(iii) This model predicts α-and β-emission properties.

(iv) This model explains certain features of nuclear fission.

(v) This model is the forerunner of the collective model of nuclear structure.

Defect. Although this model explains certain features of nuclear fission, yet it is not vary successful in describing the actual excited states as it gives too large level distances.

(d) Nuclear Shell Model. This model was proposed by Haxel, Mayer Feenberg, Nordheim and others. According to this model, the magic numbers of neutrons and protons have been interpreted as forming closed (completed) shells of neutrons or protons in the nuclei in analogy with the filling of electrons shells in atoms. These neutron and proton shells in the nucleus are independent of each other.

In this model it is assumed that a nucleus moves independently in a common mean potential due to the remaining nucleons. So this model is also called independent particle model. Assuming a spherically symmetric central field of force the Schrodinger equation can be written and the solution can be obtained.

This is the basis of shell model. Based on this solution the nucleons can be accommodated in different shells; like filling of electrons in atomic structure. Here one finds that closed shells are obtained when the proton and neutron number correspond to magic number.

The nuclear shell model compares with the electron shell model of the atom in that shells are regarded as "filled" when they contain a specific number of nucleons. A nucleus which has filled shells is more stable than one which has unfilled shells. Extra–stable nuclei are thus analogous with the inert gas atoms which have filled electron shells. The elucidation of the laws governing the numbers of nucleons in filled nuclear shells is, however, not so readily possible as is in the case of electrons. Though the Pauli exclusion principle applies, theoretical task is much more difficult because :

(i) There are two different types of nucleons concerned, the protons and neutrons, as compared with electrons only in the extra nuclear structure of the atom.

(ii) Unlike the extra nuclear electrons, nucleons are not subjected to the attraction of a central force of electrostatic origin; indeed, the nature of inter-nucleon forces is not fully understood.

Evidences for the Nuclear Shell Model. Evidence for the number of nucleons in closed shell is consequently semi - empirical and based on a study of the stability and interaction of the many nuclides known. Thus,

(1) Nuclides having closed shells would be expected to have a number of stable isotopic or isotonic forms. Tin with $Z = 50$, *i.e.*, there are 50 protons in the nucleus, has 10 stable isotopes more than any other element.

(2) The binding energies of atomic masses in atoms with the magic number nucleons total, as calculated by the Einstein mass energy equation, are higher than their neighbours.

(3) Radioactive atoms which are alpha emitters tend to attain the stable configuration associated with the magic number total.

(4) Nuclei with nucleons total just above a magic number are less stable than those with magic number total. Such nuclei may be neutron emitter, and in the process of emitting neutrons they tend to attain the magic number total. Nuclei of this type are extremely short - lived and occur only as intermediate products in processes involving artificial radioactivity.

(5) The electric quadrupole moments of nuclei exhibit sharp minima at the closed shell numbers indicating that such nuclei are nearly spherical.

(6) The asymmetry of the fission of uranium could involve the sub-structure of nuclei, which is expressed in the existence of magic numbers.

(7) A good example for the closed nuclear shell is the extreme stable $_2He^4$ and $_8O^{16}$. They have magic numbers of nucleons, *i.e.*, 2 protons + 2 neutrons in Helium and 8 protons + 8 neutrons in oxygen.

(8) The nuclides containing magic number of neutrons are more reluctant to absorb one additional neutron.

(9) The nuclei having one additional neutron above magic number are having more inclination to transfer into a nucleus with magic number by expelling out that additional neutron.

$$_8O^{17} \rightarrow {_8O^{16}} + {_0n^1}$$

(10) Alpha particle disintegration energy is found to be large if the daughter nuclide; has magic number of nucleons while it is found to be very small if the magic number is associated with the parent,

(11) Elements of nuclei containing magic number of nucleons have large number of stable isotopes.

(12) The nuclides which are more abundant in nature are found to be $_8O^{16}$,20Ca^{40}, $_{38}Sr^{88}$, $_{39}Y^{89}$, $_{40}Zr^{90}$, $_{50}Sn^{118}$, $_{56}Ba^{138}$.

(13) All the end 'product of 'the 'four radioactive series are $_{82}Pb^{206}$ $_{82}Pb^{207}$ $_{83}Pb^{208}$ $_{83}Pb^{209}$. They are associated with magic numbers.

(14) Total nuclear angular momentum can be explained by shell model of nucleus. In even - even nuclides all the protons and neutrons should pair off so as to, cancel out each other spin. Thus even - even nuclides are expected to have zero nuclear angular momentum. In even odd or odd - even nuclides the 1/2 integral spin of the single extra nucleon will exist. Hence the nucleons of this type will have 1/2 integral spin. In the case of odd-odd nuclei each has an extra neutron and an .extra proton.

(e) Collective Model. This is a combination of the above two models. Its basis is nuclear quadrupole moment which may be regarded as a measure of deviation from spherical symmetry. It is assumed that the particles within the nucleus are exerting a centrifugal force on the surface of the nucleus and as a result of which the nucleus is distorted into a non--spherical shape and the surface starts oscillating (liquid drop model). The particle now comes under non -spherical potential. Thus nuclear distortion reacts on the particles and modifies the independent particle aspect (shell model). Hence a nucleus may be considered as a shell structure which is capable of performing oscillations in shape and size. This model is useful,

(i) For predicting correctly the properties of low energy levels of nearly spherical nuclei close to closed shells, and

(ii) For clearly understanding the problems which are related with magnetic moment, nuclear quadrupole moment and isomeric transitions.

(f) Jastrow's Theory. Jastrow (1951) postulated the hard core structure of the nucleus. In this structure and inner repulsive potential acts along with an attractive force for interaction between triplet and singlet states. The radius of internal core is almost 0.6×10^{-3} cm which is approximately half of the range of nuclear force (1.25×10^{-13}cm). The hard core structure of nucleus is able to predict about the possibility of saturation character of nuclear forces. This saturation may arise due to other causes such as *zero point energy, exchange forces, core effect , etc.*

(g) Nilson's Model. This model was proposed by Nilson in 1955 and combines both the aspects. In particle concept model (shell model) the potential energy field in which' the partible moves is supposed to be of spherical symmetry. In Nilson's model individual nucleons move in such a potential field that can be deformed especially in the regions between closed shells. This distortion in potential gives rise to different pattern of filling the nucleons in quantum levels than that required in independent particle model and the description of various energy levels (Nilson states) needs new quantum numbers. This model Successfully explains the various concepts of nuclear behaviour *e.g.*, nuclear spin, magic numbers, quadruple moments, low-energy excited states, etc.

WARNER'S THEORY OF COORDINATION

Introduction : An understanding of co-ordination compounds and their properties began with the work of *Alfred Werner.* In 1899, Werner proposed an interpretation of co-ordination compounds which emphasized on the number and nature of groups attached to the metal ion. This theory replaced the older concepts of *Berzelius* (*1819*), *Graham* (*1837*), *Claus* (*1856*). *Blomstrand (1869) and Jorgensen* (*1878–1894*) and became a fundamental part of the electronic theory of valency formulated by *G. N. Lewis* (*1916*).

According to Werner,

(a) Metals possess two types of valencies, so called : (a) Principal or ionisable valency and (b) Secondary or non-ionisable valency.

(b) Primary valencies are those which a metal exercises in the formation of its simple salts. Thus primary valencies of Pt, Co, Cu and Ag in the formation of their simple salts, *e.g.*, $PtCl_4$, $CoCl_3$ and $CuSO_4$ and AgCl are 4, 3, 2 and 1 respectively.

(c) Secondary valencies are the valencies which a metal cation exercises towards a neutral molecule or an anion in the formation of complex ions. The secondary valency is also termed as the coordination number of the metal cation under consideration. Thus, co-ordination number gives the number of neutral molecules or onions which may be linked to the metal cation in the formation of its complexes. For example, cobalt (III) and platinum (IV) were recognised as having six secondary valencies (co-ordination number = 6), whereas copper (I) has four (co-ordination number = 4).

(d) Primary valencies are satisfied by negative ions whereas secondary valencies may be satisfied by either negative groups or neutral molecules.

(e) Secondary valencies are directed in space and hence such compounds are capable of exhibiting the phenomenon of isomerism.

On the basis of this theory, Werner assigned to $CoCl_3.6NH_3$ the structural formula. In this formula, size ammonia molecules are attached to the cobalt by secondary valencies where as three chlorine atoms are held to the cobalt by means of primary valencies. *Primary valencies are designated by solid lines and secondary valencies by dashed lines.*

EXPERIMENTAL EVIDENCE IN SUPPORT OF WERNER'S THEORY

A large number of complexes were studied by Werner and other workers, Werner used molar conductance values to deter-mine the number of ions per molecule thus assigning groups to the first co-ordination sphere. Chemical methods like pre-cipitation of Cl^- by $AgNO_3$ to determine the number of ionisable CF ions per molecule were also used. An interesting example is being discussed below :

Cobalt (III) ammine complexes : When an aqueous cobalt (II) chloride solution is treated with ammonia and then oxidised by air, the following six compounds are isolated :

(a) Luteo cobaltic chloride, $CoCl_3{\cdot}_6NH_3$: This is an orange yellow crystalline compounds. Its chief characteristics are :

(i) *Treatment of this compound in the solid state with sulphuric acid liberaıes all the chlorines as hydrogen chloride and leaves a sulphate. $Co_2(SO_4)_{3.}12NH_3$. Treatment of its solution with silver nitrate precipitates all the chlorines immediately. Thus, the bonding between cobalt and chlorine is purely ionic.*

(ii) *Treatment of solid with hydrochloric acid even at 100°C affects no removal of ammonia. Similarly, treatment of the solid with moist silver oxide gives a water soluble compound of composition $CoOOH.6NH_3$ which on addition of acids reforms the salts of type*

From the above experimental facts it follows that all the six ammonia molecules are unionisable and are held to the metal ion Co^{3+} in the first or inner coordination sphere whereas the ionisable Cl^- ions are in outer coordination sphere. From the above considerations, Werner formulated this complex as:

$$[Co(NH_3)_6]Cl_3$$

Conductance measurement indicates the presence of four ions, *i.e.*, one cobalt and three chloride ions (ammonia molecules are neutral). This is in accordance with the dissociation,

$$[Co(NH_3)_6\ Cl_3 \quad + 3Cl^-$$

(b) Roseo cobaltic Chloride $CoCl_3.{}_5NH_3.H_2O$: This is a pink crystalline compound with the following characteristics :

(i) With $AgNO_3$, all the chlorines are precipitated as chloride from the solution of this complex indicating their ionizable character.

(ii) The water present in compound is firmly held at room temperature and is lost only at 100°C or above, showing its presence in the inner coordination sphere.

(iii) On treating with cone. H_2SO_4 it yields a compound $Co_2(SO_4)_3$ $10NH_3.\ 5H_2O$ which on slight heating loses three water molecules, forming $Co_2(SO_4)_3\ .10NH_3.2H_2O$.

The above facts clearly indicate that water molecules and five ammonia molecules are firmly bound with cobalt whereas chlorines are loosely bound with cobalt. Thus, the formula of complex is

$$[Co(NH_3)_5.H_2O]Cl_3$$

Conductance measurements indicate the presence of four ions per molecule in very dilute solutions. This is in accord with the dissociation.

$$[Co(NH_3)_5.H_2O]Cl_3 \quad [Co(NH_3)5.H_2O]^{3+} + 3Cl^-$$

(c) Purpureo cobaltic chloride $CoCl_3.5NH_3$: It is a violet coloured compound with the following characteristics :

(i) When treated with cone. H_2SO_4, two chlorine atoms are lost as HCl, forming $Co(Cl)SO_4$.

(ii) Only two chlorines are precipitated immediately as AgCl on adding silver nitrate to its solution.

The above mentioned facts indicate the one chlorine atom is different from the rest two chlorine atoms. It means that only two of the three chlorines are ionisable as Cl^- ions.

On the basis of the above facts, Werner postulated the following formula : $[Co(NH_3)_5CL]Cl_2$

Conductance measurements indicate the presence of three ions per molecule in very dilute solutions. This is in accord with the dissociation.

$$[Co(NH_3)_5Cl]Cl_2 \quad [Co(NH_3)_5Cl]^{2+} + 2Cl^-$$

(d and e) Praseo-cobaltic chloride, violeo cobaltic chloride. $CoCl_3.4NH_3$: Two related com-pounds, one violet in colour (violeo-cobaltic chloride) and the other green (praseo-cobaltic chloride), have been prepared. Both have the same composition $CoCl_3.4NH_3$ but the chemical behaviour of these two compounds is quite different.

The conductances of solutions of these compounds show the presence of only two ions and from them only one chlorine is immediately precipitated by silver nitrate. Thus, from structural point of view their composition may be written as, $[Co(NH_3)_4Cl_2]Cl$

However, there were distinct differences in chemical behaviour between the two. Thus, although only one third chlorine gets precipitated immediately on the addition of silver nitrate, second and third choleric atoms per molecule get also precipitated slowly and after long standing and no boiling. Werner explained this by assuming that the chlorines of inner co-ordination sphere are slowly replaced by water molecules :

$$[Co(NH_3)_4Cl_2]^+ + H_2O \quad [Co(NH_3)_4H_2O.Cl]^{2+} + Cl^-$$

$$[Co(NH_3)_4H_2O.Cl]^{2+} + H_2O \quad [Co(NH_3)_4.2H_2O]^{3+} + Cl^-$$

However, the rate of precipitation of these chlorines in these compounds is different.

Werner explained the above facts by assuming that praseo-cobaltic chloride is trans- type whereas violeo-cobaltic chloride is cis- type.

(f) $CoCl_3.3NH_3$: This is a blue green compound having the following characteristics :

(i) *It does not give any precipitate immediately on addition of silver nitrate to its solution (absence of ionisable chlorine).*

(ii) *Its dilute solutions did not conduct electricity at all, showing the absence of any ions in solution.*

From the above facts Werner came to the important conclusion that all the groups are linked to the central metal ion by unionisable secondary valencies and gave the following formula : $[Co(NH_3)_3Cl_3]$

Werner applied his theory to the above ammonia compounds of cobalt (III) chloride. The compounds may be formulated as :

In all the compounds discussed above, the primary valencies are designated by solid lines and secondary valencies by dotted lines. In each case, satisfaction of six secondary valencies is essential. Werner's theory is successful in

(i) Predicting the correct number of ions in each case.

(ii) Firm association of chlorine with cobalt in the last three cases and

(iii) The existence of two forms of $CoCl_3.4NH_3$ is explained. It is due to difference in three dimensional arrangement of four ammonias and two chlorines about the central cobalt (III) ion.

In order to differentiate, materials held by secondary valencies are enclosed in square brackets when writing formulae whereas materials held by primary valencies outside the brackets. Thus the compounds just considered are written as.

$$[Co(NH_3)_6]Cl_3), [Co(NH_3)_5(H_2O)] Cl_3, [Co(NH_3)_5Cl]Cl_2$$
$$[Co(NH_3)_4Cl_2] Cl \text{ and } [Co(NH_3)_3Cl_3]$$

Criticism of Werner's Theory : One must appreciate the remarkable work done by Werner. He did this work 70 years ago and inspite of advances in theory and the increase in the number of co-ordination compounds, his postulates are as valid today as when they were presented over 70 years ago.

2. Electronic Interpretation of Co-ordination

Sidgwick and Lowry gave an electronic interpretation of Werners theory.

According to Sidgwick, *Werner's primary valencies were regarded as amounting to electron transfer and the secondary or nonionic valencies to electron pair sharing.*

An important point is to be noted that *all neutral molecules or onions which were capable of being coordinated to central metal ion have atoms with at least one unshared lone pair of electrons in their valency shells.* Thus, the linkage of these co-ordinating groups to metal ion by secondary valencies arises due to the donation of the electron pairs to the central metal by these groups. *Bonds of this type are termed as co-ordinate bonds. These bonds have been indicated by arrows.* For example the structure of $[Co(NH_3)_6]^{1+}$ ion.

Effective Atomic Number (EAN): Stability of the complex compounds is explained by the fact that central metal atom acquires the effective atomic number of the next inert gas. *The effective atomic* *number is defined as the number of electrons of the central metal in a complex including those gained by bonding.*

In most of the cases the effective atomic number is the same as the atomic number of the next inert gas. However, there are a few cases, where the effective atomic number is more or less by one or two units as compared to next inert gas. EAN of some metals in certain complexes. It is evident that tendency to attain an inert gas configuration is a significant factor but not a necessary condition for complex formation. Some complexes of chromium and platinum are equally stable, yet the effective atomic number is short of inert gas configuration. For complex for-mation, it is essential to produce a symmetrical structure, *i.e.*, tetrahedral, square planar, octahedral, without any consideration of the number of electrons involved.

Objection. Sidgwick's electronic interpretation of co-ordination is not entirely satisfactory. Some objection are :

(i) The donation of electron pairs to a central cation would produce an improbable accumulation of negative charge on this ion. In the case of $[Co[NH_3)_6]^{3+}$ the donation of electron pairs from the ammonia molecules to the cobaltic ion would result an accumulation of negative charge on cobaltic ion. In this way cobaltic ion would become negatively charged with respect to the ammonia molecules.

(ii) Another difficulty presented by this concept is the fact that the electron pair, available for donation in water, ammonia and many other neutral molecules, is $2s^2$ pair. The electrons have no bonding characteristics and to excite them to a higher energy level would require more energy than is usually available in bond forma-tion.

3. Pauling's Theory

In 1931, Pauling introduced a new theory which was based on the revolutionary idea of hybridisation. His theory was able to account for and even predict the geometrical shapes of many complexes. According to him.

(i) Strong bonds are not formed by pure s-or pure p- or pure d-orbitals.

(ii) Strong bonds are generally formed by hybridised orbitals formed by a suitable combination of orbitals. For example, the combination of s-orbital and p orbitals gives four equivalent orbitals of equal energy known as hybrids. This type of combination is known as sp^3 hybridisa-tion.

In complexes, other types of hybridisation like d^2sp^3, sp^3d^2 and dsp^2 are more common.

(iii) In transition metal atom, d-orbitals are, partially filled up and hence it is the number of d-electrons in the central atom (forming complex) that determines the possible modes of hybridization and the shapes of complex.

Pauling's theory provides a suggestive explanation of the experimental facts and, in particular, of the magnetic susceptibilities of co-ordination compounds. Pauling's theory also differentiated between com-plex ions in which the bonding could be considered as ionic and those in which it was purely covalent.

Let us apply this theory to various examples;

Octahedral complexes : Octahedral complexes result from either sp^3d^2 or $d^2\,sp^3$ hybridisation as in the following examples.

(a) Ferricyanide ion, $[Fe(CNH)_6]3^-$: The structures of the Fe atom and the Fe^{3+} ion are given at (i) and (ii).

The five unpaired electrons in the Fe^{3+} give a calculated magnetic moment of 5.91 magneton. The magnetic moment for the ion, however, is 2.3 suggesting, perhaps, one unpaired electron.

In order to explain the magnetic moment of , Pauling suggested that the unpaired *3d* electrons in the Fe^{3+} undergo pairing to give the arrangement as shown in (ii). The resulting complex would have one unpaired electron, and the d^2sp^3 orbitals are available for hybridisation. The electrons, to occupy the d^2sp^3 hybrid orbitals, would be provided by the lone pairs from each of the six cyanide ion ligands. This means that the bonds between the ligands and the central ion are being coordinated linkages.

Complexes like $|Fe(CN)_6]^{4-}$ are known as *inner-orbital complexes* because the *d-orbitals* involved are from a lower shell than the *s* and p-orbitals.

(b) Ferrocyanide ion $[Fe(CN)_6]^{4-}$: In the ferrocyanide ion there is one more electron than in the ferricyanide ion. This extra electron pairs with the unpaired 3d-electron in (iii), to give the structur. The ferrocyanide ion has, therefore, no unpaired electrons and would be expected to be diamagnetic, this is found to be so. It is octahedral with d^2sp^3 hybridisation. Other examples resembling structures like the ferri , or cyanide ions include , and .

(c) Ferrifluoride ion $[FeF_6]^{3-}$: The $[FeF_6]^{3}$ is octahedral like the ferri- and ferrocyanide ions. However, it has a magnetic moment of 6,0 which is in close agreement with the value of 5.9 – a value which would be expected from a structure with five unpaired electrons.

In order to explain this, it is assumed that no pairing of electrons in Fe^{3+} ions takes place and sp^3d^2 orbitals are used for hybridisation. The is an example of an outer orbital complex because the d-orbitals involved in the hybridisa-tion are from a higher shell than the s- and p-orbitals.

B. Tetrahedral Complexes : Tetrahedral complexes result from sp^3 hybridisation. Complexes of Zn^{2+} are tetrahedral because they involve sp^3 hybrid bonds. The electron arrangements in the Zn atom and Zn^{2+} ion are shown at (i) and (ii) below :

In the Zn^{2+} ion, all the $3d$ orbitals are fully occupied and therefore, they cannot take part in bond formation. However, sp^3 hybridisation takes place and four sp^3 hybrid orbitals are formed. The electrons to occupy sp^3 hybrid orbitals would be provided by the lone pairs from each of the four ammonia molecules in . Thus, will be tetrahedral. As there are no unpaired electrons, will also be diamagnetic.

Other similar tetrahedral complexes are formed by Cd^{2+}, Hg^{2+}, Cu^{+} and Ag^{+} ions; all of which have the same structure as Zn^{2+}.

C. Square planar complexes : Square planar complexes result from dsp^2 hybridisation. This type of hybridisation involves mixing of one *d-*, one *s-* and two $2p$ orbitals, resulting in the formation of four equivalent hybrid orbitals which are planar and mutually at right angles pointing towards the four corners of a square. The d–orbital involved is the $dx^2 - y^2$ orbital in the shell lower than that of the *s-* and p-orbitals.

Square planar complexes tend to be formed when the central ions have only one d-orbital available in the lower shell.

(a) Nickelocyanide ion : The electronic arrangements for the Ni atom and the Ni^{2+} ion are shown below :

If the two unpaired $3d$ electrons couple, the structure will be as given in (iii) so that dsp^{2-} hybrid bonds can be formed. All the electrons are paired so that the ion is diamagnetic. Other similar examples are, etc.

(b) Cuprammonium ion. : It has a coplanar square structure and is paramagnetic with a magnetic moment of 1.8. A square planar structure is obtained by dsp^2 hybridisation, keeping one electron unpaired. Such facts are accounted for on the basis of the structures shown at (i), (ii) and (iii).

Objections to the Shape of Cuprammonium Ion

(a) The necessity to promote one electron from a $3d$ into a $4p$ level is not satisfactory. The 4p electrons would be expected to be easily lost which would mean that the complex could be easily oxidised, but this is not so.

(b) Placement of the odd election in the $3d$ level as in (iv) would suggest sp^3 hybridisa-tion with a resulting nonexistent tetrahedral structure.

Objections to Pauling's Theory : Pauling's theory explained the chemistry of co-ordination com-pounds satisfactorily but failed to explain some of the aspects :

(a) The peculiar behaviour of ions of divalent nickel, palladium, platinum and trivalent gold which apparently do not form the expected 5–co-ordination complexes.

(b) Pauling's theory proposed a form of hybridisation (dsp^2) suitable for square-planar structures but this does not explain why this structure is preferred to other possible stereochemical configurations, such as tetrahedral or 5-co-ordinated trigonal bipyramid which does, in fact, occur in iron carbonyl.

(c) Pauling's ideas offered a little or no explanation to the absorption spectra of complexes or of their relative stabilities or reactivities.

(d) It is well known that both the colour and magnetic moment of transition metal complexes are due to their possessing d-orbital electrons. It means that there should be a quantitative connection between spectra and magnetic moment. Unfortunately this connection is not revealed by Pauling's theory.

(e) It cannot explain why certain complexes are more labile than the others.

Labile complexes are those in which one ligand can be easily displaced by another ligand. On the other hand, inert complexes are those in which displacement of ligands is slow.

(f) It does not take into consideration the splitting of d–energy levels.

(g) It does not account for the relative energies of different structures.

4. Electroneutrality Principle

Earlier while discussing the Sidgwick's interpretation of Werner's compounds we have shown that coordinate bonds are formed by the donation of a pair of electrons by each ligand atom to metal ion. Once the bond is formed there is no ditinction between a coordinate bond or a covalent bond for in both the cases the electrons are equally shared. Thus, in $[Co(NH_3)_6]^{3+}$ complex, out of 12 electrons shared between cobalt (III) and 6N atoms of ammonia molecule, the metal gains 6 electrons and so the formal charge changes from +3 to –3. Thus, the formation of coordinate bonds implies the accumulation of negative charge on the metal atom.

But Pauling has shown that such a situation need not arise which he expressed in his *electroneutrality principle.* According to this principle *the electrons are distributed in a molecule in such a way as to make the residual charge on each of atom zero or very likely zero except that hydrogen and the most positive metals can acquire partial positive charge and the most electro-negative atoms can acquire partial negative charge.* Thus obviously it implies that the M–L bond is not perfectly covalent. In other words, the bond is partially ionic and that the electron pair is attached more strongly to the ligand atom. Thus, if it is supposed that the M–L bond is 50 percent ionic, then the metal would be exactly neutral and the charge on each ligand atom would be +1/2. The formation and stability of aquo and ammonia complexes can be explained on the basis of ion dipole electrostatic attraction but for ligands having negative character electroneutrality principle has to be invoked to explain their stability.

In addition Pauling has also suggested that accumulation of negative charge can be avoided if it is assumed that in addition to o bond, the central metal atom also forms n bonds by the back donation of electrons to the ligand atoms. Thus, the formation of

M L

π bond results in the decrease of negative charges on the metal atom. The *n* bond formation may take place in two ways :

(i) ***dπ–pπ. bonding*** : *The first atom in a group which acts as donor atoms, (e.g., N,C in NO_2^- and CN^- respectively) do not have d-orbitals available for bonding. So the electrons donated by the filled dn orbitals (dxy, or dyl orbitals) are accommodated in the vacant p orbitals of the ligand atom. Thus, there is an effective dπ–pπ overlapping between metal and ligand orbitals.*

(ii) ***dπ-dπ bonding*** : *In this type of bonding the back donated electrons from dπ orbitals of the metal are accommodated in the dπ orbitals of the donor atoms. Such donor atoms, naturally will have vacant dn orbitals and belong to the second short period or of the later periods e.g., P, S, As in PR_3, SR_2 and AsR_3, respectively.*

The formation of π bonding not only avoids the accumulation of negative charges on the metal atom but also strengthens the σ bond which explains the stability of many transition metals complexes with very poor electron pair donor atoms, such as CO, PX_2, PR_2, SR_2, and

5. Crystal Field Theory

Introduction : This theory which was first introduced by Bethe and Van Vleck and applied mainly to ionic crystals, and is, therefore, called **Crystal Field Theory** (1956). Later **Orgel** was largely respon-sible for developing its general chemical aspects. The crystal field theory is mainly concerned with the interaction of d-orbitals of central metal with the surrounding ligands that produce crystal field effects. The crystal field theory is based on the following assumptions:

(i) The metal ion and ligands act as point charges and the interaction between them is purely electrostatic, *i.e.*, metal-ligand bonds are 100 per cent ionic.

(ii) There is no interaction between metal orbitals and ligand orbitals.

(iii) The d–orbitals on the metal which were all of the same energy, *i.e.* degenerate in a free atom, have their degeneracy destroyed by the ligands when a complex is formed.

Salient Features of Crystal Field Theory

The various postulate are summarised below

(a) A complex is considered to be a combination of central metal ion surrounded by various ligands.

(b) The ligands in the complexes to be considered are either negatively charged ions *e.g.*, F and CN^- or neutral molecules, *e.g.* H_2O and NH_2, which have a dipole such that the negative ion is closest to the central ion.

(c) The attraction between central metal and ligand in a complex is regarded as purely electrostatic. The interaction between the electrons of the cation and those of the ligands is purely repulsive. It is these repulsive forces that are responsible for causing the splitting of the d-orbitals of the metal cation into two groups, t_{2g} and *eg*. The splitting depends on whether the ligands are arranged in an octahedral, tetrahedral or squareplanar way around the central ion. This effect is known as *crystal field splitting.*

(d) The electrons of the metal ion are repelled by the negative field of the ligands and therefore the metal electrons will occupy those d-orbitals which have their lobes farthest way from the direction of the ligand.

(e) In this theory, the covalent character of bond between metal and ligand is not taken into account.

(f) This theory does not permit the electrons to enter the metal orbitals. It means that it does not consider any orbital overlap.

(g) The number of ligands and their arrangement around the central ion will determine the crystal field.

Crystal field splitting of d-orbitals : In an isolated gaseous metal ion, all the five d orbital are degenerate, *i.e.*, have the same energy. The electrons which occupy such orbitals, do so according to the Hund's rule of maximum multiplicity.

Crystal field theory is basically concerned with the effect of different arrangement of surrounding ligands on the energy of the d-orbitals, and the subdivision of the orbitals into two groups, t_2g and e_g.

On the approach of the ligands, the electrons in the d-orbital of the central ion are repelled by the lone pairs of the ligands. This repulsion will raise the energy level of the d-orbitals. All the ligands approaching the energy of each orbital will increase by the same amount. In other words they will remain degenerate, although they will have now higher energy than before.

But actually the d-orbitals differ in their orientation. Thus the energy of the orbitals lying in the direction of the ligands is raised to a larger extent than that of the orbitals lying in between the ligands. It means that the five degenerate d-orbitals of the metal ion will split up into two sets of orbitals having different energies This splitting of five degenerate d-orbitals of the metal ion under the influence of approaching ligands, into two sets of orbitals having different energies, is called *crystal field splitting or energy level splitting.*

The ligand field splitting depends on whether the ligands are arranged in an *octahedral, tetrahedral or square planar* way around the central ion. The magnitude of the crystal field splitting of the d-orbitals energies may be represented by the symbol Δ. It is used to denote energy difference (the total energy difference may be regarded as made up of the energy lost or released in the lower levels together with the energy gained in the upper levels).

The crystal field splitting energy, Δ, is frequently measured in terms of a parameter, Dq. The mag-nitude of splitting is arbitrarily set at 10 Dq, *i.e.*, 2 Dq for each unpaired d–electron.

CRYSTAL FIELD SPLITTING BY DIFFERENT GEOMETRICALLY ARRANGEMENTS OF LIGANDS

The way in which the degeneracy of d-orbitals is resolved is different in octahedral, tetrahedral and square planar fields. We shall discuss separate-ly the splitting in different fields.

(A) Octahedral complexes : In the octahedral complex, the six ligands are arranged octahedral around a central metal ion.

In such an arrangement as the dx2--y2 and *dz*2 orbitals lie along the *x, y* and *z* axes and so point directly towards the ligands, they experience much more repulsion than the remaining d-orbitals (*dxy*, *dyz* and *dzx*) which are directed in between the or, y and z axes. Consequently, the energy of *dx*2 – *y*2 and dz2 orbitals is increased much more in comparison to other d-orbitals. Thus, the d-subshell is splitted up into two degenerate sets, one consisting of more stable (lower energy) orbitals *dxy, dyz* and *dzx* and the other, less stable higher energy) orbitals dx2 – y2 and dz2. These are commonly referred to as t2g and eg or dÎ and dg sets of orbitals respectively.

Degeneracy of all the five d–orbitals in the isolated central ion is represented by state I. The state II represents degeneracy (hypothetical) of all the five orbitals at a higher energy level. The state III represents *crystal field splitting.* The crystal field splitting is thus the energy difference between t_{2g} and e_g orbitals. It is denoted by the symbol Δ_0 or 10 Dq. (The subscript 0 indicated octahedral complexes).

The geometry of an octahedral system shows that the energy of the t_{2g} orbitals is 2.5 Δ_0 or 0.4 Δ_0 or 4 Dq less than that of the hypothetical degenerated d orbitals which results in crystal field splitting is ignored. The energy of the e_g orbitals is 3/5 Δ_0 or 0.6 Δ_0 or 6 Dq above that of the hypothetical degenerate d–orbitals.

An electron prefers to move into an orbital of a lower energy. It is obvious that if an octahedral complex contains one d-electron, this electron will occupy one of the t_2g orbitals. The t_2g orbital has an energy 4 Dq less than that of the degenerate orbitals. Thus the complex will be 4 Dq more stable than predicted by the purely electrostatic theory which does not recognise any splitting of energy levels. This value, 4 Dq, is called the *crystal field stabilising energy* (C.F.E) of the complex under discussion.

If for each electron entering into a tag orbital, the crystal field stabilising energy is 4 Dq. then for each electron entering into an e_g orbital, the *destabilising energy is* 6 Dq. In other words, the stabilization energy assigned for an electron in e_g orbital is – 6 Dq. Working on this basis, it is possible to calculate crystal field stabilization energies for various metal ions in octahedral complexes.

Distribution of electrons in *d* orbitals and CFSE of octahedral complexes : The distribution of electrons in the *d* orbitals is based upon the following rules :

(i) *Electrons prefer to occupy the orbitals of lowest energy.*

(ii) *Electrons obey Hund's rule, i.e., electrons pairing will not take place until all the available orbitals of a given set contain one electron each.*

(iii) *If the energy to place an electron in the upper e_g level is greater than the energy to pair electrons (pairing energy* = *P) in the lower* t_2g *level* (10 Dq < P), *then the electrons will pair up in* t_2g *orbitals.*

(iv) *If the energy to place an electron in the upper eg level is less than the pairing energy in the lower* t_2g *orbital* (10 Dq < P), *then the electrons will occupy e_g orbitals.*

Weak and strong field cases. Ligands which cause only a small degree of splitting of *d* orbitals, *i.e.,* weak field, are called *weak field ligands* and those which cause a large splitting. *i.e.,* strong field, are called *strong field ligands.* The magnitude of 10 *Dq* and hence the distribution of the electron in *d* orbitals are very much influenced by the strength of the ligand field. The CF splitting ability of the ligands decreases in the following order :

$> NH_3 > H_2O > F^- >$

In a weak field, 10 Dq is generally less than pairing energy, *i.e.,* 10 Dq < P. It means that electrons obey Hund's rule and remain unpaired. But in the case of a strong field, 10 Dq > P; it follows that electrons pair up in the lower energy orbitals. We shall now consider the distribution of *d* electrons in octahedral complexes, in the presence of weak field and strong field ligands.

In octahedral complexes with d^1 configuration, the *d* electron will occupy the lower t_{2g} orbitals irrespective of the field strength and the crystal field stabilisation energy is – 4Dq. For d^2 configuration, the two electrons will occupy the lower two t_{2g} orbitals and CFSE is – 4Dq × 2 = 8 Dq. Similarly for d^3 configuration CFSE is – 12Dq and the t_{2g} level is just half-filled. Thus complexes with d^1, d^2 and d^3 con-figuration, irrespective of the ligand field strength, remain paramagnetic, with one, two and three unpaired electrons respectively.

For d^4 configuration, two possibilities will arise, that is, the fourth electron can either enter the upper e_g level or pair up in the lower t_{2g} level. Which of these two arrangements occurs depends upon the values of 10 Dq and P, which in turn depends upon the strength of the ligands. In a *weak field,* 10 Dq < P, hence the fourth electron prefers to enter one of the e_g orbitals rather than pairing in t_{2g} orbitals. This way, it loses somewhat less energy in the form of CFSE by entering the destablised e_g level. The net CFSE energy is :

$$\text{CFSE} = (3 \times -4\text{ Dq}) + 1 \times (+6\text{Dq}) = -6\text{Dq}$$

In a *strong field* where 10 Dq > P, the fourth electron prefers to pair up in one of the three lower t_{2g} orbitals rather than entering the upper e_g orbitals. CFSE for this type of complex is (4 × – 4Dq + P) because all the four electrons enter t2g orbitals and some energy is used to pair up the electron.

Thus, for d^4 configuration the electronic arrangement' differs in weak and strong fields. In weak field there are four unpaired electrons while in the strong ligand field, there are only two unpaired electrons. These two arrangements thus differ in the number of unpaired electrons. The arrangement with maximum unpaired electrons is called *'high spin'* or *'spin free'* and the one with minimum number of unpaired electrons is called *low spin or spin paired* arrangement.

For d^5 configuration, in a weak field three electrons occupy the t_{2g} orbitals and the other two in two e_g orbitals (one in each). CFSE is zero because the presence of two electrons in the unfavourable e_g level exactly balances, the stabilisation from the three electrons in the t_2g level. In other words, the half filled shell (d^5) is physically symmetrically and no stabilisation can occur through the application of external ligand field. In case of *strong field* the fifth electron will pair up in the t_{2g} orbital.

$$\therefore \qquad \text{CFSE} = -20\text{ Dq} + 2\text{P}$$

For d^6 configuration in a weak field, the sixth electron will pair up in the lower t_{2g} orbitals because all the five d-orbitals are already half-filled CFSE = – 4Dq + P. In case of *strong field,* the sixth electron will pair up in the t_{2g} orbitals. Thus, all the three t_2g orbitals will have paired electrons.

$$\therefore \qquad \text{CFSE} = -24\text{ Dq} + 3\text{P}$$

Similarly, *for* d^7, d^8 and d^9 and d^{10} configurations, the d–electrons of the metal can be distributed and their CFSE can be calculated.

It is clear that for d^1, d^2,d^3, and d^8, d^9, d^{10} configura-tions, the distribution of electrons and CFSE are the same irrespective of the strength of the ligand field. But for configurations d^4 to d^7, two arrangements (high-spin and low- spin) are possible depending upon the strength of the ligand field.

B. Tetrahedral complexes : In tetrahedral complexes, the four ligands occupy the alternate corners of a cube, in the centre of which is placed the metal cation. The lobes of the dxy, dyz and dzx orbitals point towards the middle of the edge of the cube while those of the $dx^2 - y^2$ and dz^2 lie on the bisectors of the angle between pairs of ligands and so pass through the centre of cube faces.

Obviously, none of the d-orbitals point directly towards the ligands. However, the calculations reveal that the *dxy dyz* and *dxz* (termed t_{2g}) are closer to ligands than $dx^2 - y^2$ and dz^2 (termed e_g). In the formation of a regular tetarahedral structure in a complex, the approach of ligands raises the energy of both sets of orbitals, but since the t_{2g} orbitals correspond more closely to the position of the ligands, their energy increases most as compared to e_g. Thus, e_g orbitals (lower energy) are filled first. This is opposite to what happens in an octahedral complexes. For maintaining the average constant, t_{2g} orbitals are raised to 4 Dq while e_g orbitals are lowered by 6 Dq from the barycenter. Thus, *the energy level scheme for the tetrahedral symmetry is exactly the reverse of that for octahedral symmetry.*

It has been observed that the energy separation between the e_g and t_2g levels Δ_1, in tetrahedral symmetry is smaller than in octahedral symmetry Δ_0 (10 Dq). This observation is as expected because the d-orbitals are not directly affected by the ligands and moreover the number of ligands is smaller. For equivalent ligand charges and distances :

$$\Delta_t = - \ D^\circ$$

where the minus sign means the reversal of the orbital levels. Because of the low value of Dq in tetrahedral symmetry *crystal field favours the formation of octahedral complexes than tetrahedral complexes.*

Low spin (strong field) tetrahedral complexes have not yet been obtained, probably because of crystal field stabilisation energies are never sufficient to bring out spin pairing Thus *only high spin* or *weak field* cases are to be considered for tetrahedral complexes. This simplifies the treatment of electron configuration and CFSE. In a weak field, pairing energy (P) is more than Δ_t. Hence, electrons enter even the higher levels and t_{2g} remain unpaired until the sixth electron forces pairing. For example in a complex with d^4 con-figuration two electrons occupy the lower e_g orbitals and two occupy the higher t_{2g} orbitals, all the four electrons remaining unpaired.

Hence, CFSE = – 4Dq *i.e.*,

$$(2 \times -6Dq) + (2 \times 4Dq) = -4Dq$$

The distribution of electrons and the crystal field stabilisation energies for d^1 to d^{10} configuration are given in Table 12.10.

(c) Square planar complexes : In the square planar complexes due to the absence of the ligand groups along the z-axis, the dz^2 orbitals becomes most stable and $dx^2 - y^2$ least stable.

However, most of the square planar com-plexes are such, where the two ligands along the z-axis are not totally removed but placed at a longer distance than the ligands along the x and y axes (*e.g.*, Cu^{2+} and Ni^{2+} complexes). In such cases the *dz* orbital though becomes more stable than $dx^2 - y^2$ and d_{xy} orbitals, it remains less stable in comparison to *dxz* and *dyz* orbitals, which still form a degenerate set. The sequence of orbitals in the increasing order of energy would then be *dxz* and *dyz*, dz^2, *dxy* and $dx^2 - y^2$.

The difference in the energy levels, created within the *d* subshell by the crystal field, are small and the electrons from lower levels can be excited to the higher levels, usually by the absorption of visible light. This phenomenon of easy *"electron Transition"* is responsible for the complexes ap-pearing coloured.

Factors Affecting D

The magnitude of Δ or 10Dq is influenced by a number of factors such as :

(a) Geometry of the complex : In octahedral complexes, the splitting of *d* orbital has been found to be more than twice than in tetrahedral complexes, for the same metal ion and ligands. This difference in Δ value is due to the following facts.

(i) *In octahedral complexes six ligands are involved whereas in tetrahedral complexes only four ligands are involved. This results in* 33 *per cent decrease in the field strength, provided the other factors do not change.*

(ii) *In octahedral complexes ligands are positioned exactly in the direction of and* $dx^2 - y^2$ *orbitals whereas in tetrahedral complexes, the ligands are not positioned at any of the d-orbitals and thus have more influence on the* t_{2g} *orbitals than on the* e_g *orbitals. However, in case of square planar complexes, the degree of splitting is more than in a tetrahedral field.*

(b) Nature of the ligands : If we compare the absorption spectra of various transition metal com-plexes, we observe that the position of absorption spectra and hence the value of Δ or 10 Dq for any given metal ion depends upon the nature of the ligands attached. In order to illustrate this point, we have given two values of Δ_0 or 10Dq for a number of Cr^{3+} complexes with different ligands.

From the above table we can draw the important conclusion that the cyanide ligand produces more splitting and is therefore a strong ligand whereas chloride ligand produces less splitting and is therefore a weak ligand. Similarly, we can compare the various ligands, and then arrange them in the order of their strength, in the form of a series called spectrochemical series.

$$I^- < Br^- < S^{2-} < SCN^- < Cl^- < NO^{3-} < F < OH^-$$
$$< H_2O < NCS^-, NH_3 < en < dipy < NO_2^- < CN^- < CO.$$

(c) Charge on the metal ion : As earlier stated that crystal field theory is based on electrostatic model. It means that the ionic charge present on the central metal has a direct effect on the magnitude of 10Dq. A generalisation is that a metal with higher charge produces more splitting than an ion with lower charge. For example, Δ_0 or 10 Dq for hexa-aquo complexes of Cr^{2+} and Cr^{3+} are 39.7 Kcal mol^{-1} and 51.94 Real mol^{-1}.

(d) Position of the metal in transition series : This in an important factor and the magnitude of crystal field splitting depends upon the position of metal in transition series, *i.e.,* whether it belongs to first, second or third series. In general, the value of Δ_0 or 10 Dq increases on descending a group of transition element, *i.e.,* from first to third transition series.

Measurement of Δ or 10 Dq

The splitting of d-orbitals by a ligand field into two sets with different energies provides a possible electronic energy change within a complex. An electron in an orbital of lower energy might be excited into an orbital of higher energy by the absorption of radiation. It so happens that the radiation necessary for such excitation lies in the visible region of light and this explains why many complexes are coloured.

Study of absorption spectra of com-plexes enables Δ values to be obtained. A band or bands, in the absorption spectra, can be related to the excitation of an electron or electrons, between d-orbitals splitted by a ligand field. The ability to interpret such spectroscopic evidence is one of the advantages of crystal field theory. The situation in the $[Ti\,(H_2O)_6]^{3+}$ ion, hexaquotitanium (III) ion, is particularly simple for there is only one *d* electron (d^1) involved. In a ground state, it will occupy one of the t_{2g} (d_ϵ) orbitals but absorption of radiation will excite it into one the e_4 ($d\lambda$) orbitals. The absorption spectrum of $[Ti\,(H_2O_6]^{3+}$ shows a strong band corresponding to wavelength of 5000A°. This represents an energy change of 57.15 Kcal/mole which must be the Δ value for this complex. Light of wavelength 5000A° is green. If such a light is absorbed, the transmitted light appears blue and red, *i.e.,* purple, and $[Ti(H_2O)_6]^{+3}$

ion is in fact, purple. Table 1.15 gives a few of the values of Δ for the various transition metal ions and ligands. While value of Δ is approximately constant for ions of a given charge with the same ligand, changing the ligand does change A and thus alters the absorption spectrum associated with the metal ion. It is this change in the crystal field splitting of the d–orbital energies that is responsible for the colour change that occurs when one ligand is replaced with an other.

APPLICATIONS OF CRYSTAL FIELD THEORY

The various applications are as follows :

1. Colour of Transition Metal Complexes : Most of the complexes of transition metals give absorption bands in the visible region. It means that they are coloured. According to the crystal field theory, the colour is due to the *d-d* transitions between the *eg* and t_{2g} electrons.

The energy difference between the e_g and t_2g electrons is so small that even absorption of low radiation in the visible light may cause excitation from a lower to a higher *d* level. As a result of the absorption of some selected wave lengths of visible light, the complexes appear coloured. For example, the complex $[Ti(H_2O)_6]^+$ absorbs green and yellow light from the white light, it means that the complex will have blue and red colour *i.e.,* purple colour.

The colour intensity of a transition metal complex depends upon the magnitude of Δ. For example, the value of Δ for $[Co(NH_3)_6]^{2+}$ is very large. It means that it requires relatively large excitation energy which it gets from the blue portions of the visible light. Its aqueous solution, therefore, appears red (white = blue + red).

The magnitude of Δ depends upon electronic configuration, size and oxidation state of metal ion, nature of ligand and geometry of the complex.

2. Magnetic Properties of Complexes: The crystal field theory is successful in calculating the number of unpaired electrons in high spin and low spin complexes. With the help of number of unpaired electrons, one can find the value of 'spin only moment' by applying the formula:

$$\mu_s = \text{B.M.}$$

From a knowledge of unpaired electrons and spin only moment, it is possible to know two important aspects of complexes.

(i) The valence state of the metal ion in an octahedral complex can be calculated.

(ii) If there are more than 3d electrons, the nature of bonding in a complex may be known.

These two points may be understood by undergoing the following examples.

(i) The magnetic moment of $Hg[Co(CNS)_4]$ is 4.5 B. M. From this value, it is evident that cobalt is in the bivalent state ($Co^{2+} \rightarrow d^7$) and the bonding belongs to spin-free type cor-responding to

n = 3.

(ii) The magnetic moment of $[Fe\,(diph)_3]\,(ClO_4)_3$ is 2–2 B.M. From this value, it is evident that iron is in the trivalent state ($Fe^{3+} \rightarrow d^5$) and the bonding belongs to spin paired type which corresponds to n = 1.

The values of magnetic moments also help to decide whether a given octahedral complex is a high-spin or a low-spin complex. If the value of μ_{exp} value for the complex is less than the μ, value for the metal ion, the complex belongs to low-spin type otherwise it belongs to high-spin type.

3. Relative liabilities of Complexes : The crystal field theory gives useful information about relative liabilities of certain complexes and also provides information about the mechanism of their substitution reactions. For example,

(i) From the crystal field stabilisation energy calculations of strong field octahedral complexes, It is evident that the speed of d^0, d^1 and d^2 complexes will be greater than the corresponding d^3, d^4, d^5 and d^6 complexes. This speed is independent of $S_N{}^1$ and $S_N{}^2$ mechanism.

(ii) CFSE calculations reveal a decrease in the order of reactivity of a, d^5, d^4, d^3, d^6 complexes.

(iii) When CFSE calculations are made for weak-field octahedral complexes, it follows front these values that complexes of d^n configuration except d^3 and d^8 should lead to fast reactions irrespective of $S_N{}^1$ and $S_N{}^2$ mechanism.

(iv) CFSE calculations also predicted that d^8 complex (Ni complex) should be inert.

All the above conclusions are in full agreement with the experimental evidences.

4. Stabilisation of Oxidation States : The crystal field theory explains why certain oxidation states are preferentially stabilised by coordination to specific ligands. For example, water stabilises Co(II) whereas ammonia stabilises cobalt (III). This is evident from the following CFSE data :

5. Stereochemistry of Complexes : The crystal field theory explains that copper (II) forms square planar complexes than tetrahedral or octahedral complexes The reason for this is that copper (II) possesses a much higher CFSE. (12.28Dq) in a square planar configuration than octahedral (CFSE = 6Dq) or tetrahedral configuration (CFSE= – 2.56Dq).

The CFSE also explains that the square planar complexes of 4d and 5d metal ions form more readily than *3d* ions.

6. Distorted Octahedral Complexes : In an octahedral complex, six ligand molecules are arranged around the central metal ion. When the bond lengths are the same for all six ligands, the octahahedral is said to have a regular shape. If the bond lengths are not same in all six ligands, the shape of octahedral complex is distorted. This change in shape is known as *distortion* and the complex is called *distorted octahedral com*

In 1937 Jahn–Teller theorem was put forward to explain why certain six coordinated complexes possess distorted octahedral geometry.

According to this theorem,

(i) An octahedral complex is said to have a regular shape if the d orbitals (both t_{2g} and e_g sets) are occupied symmetrically.

(ii) When d-orbitals of the central metal ion of an octahedral complex possess t_{2g} orbitals as unsymmetrical orbitals, there occurs slight distortion from the regular octahedron. This distortion is due to unevenly filling in an octahedral complex because inese orbitals do not point towards the ligands. The slight distortion may occur in such cases when the central metal ion contains 1, 2, 4 or 5 electrons.

(iii) Strong distortion occurs when the e_g orbitals of an octahedral complex are unsymmetrically filled because these orbitals do not point towards the ligands. Due to this strong distortion the octahedral shapes of complexes may change to tetragonal and even to square planar complexes. The high-spin complexes of d^4 and d^9 and low spin complexes of d^7, d^8 and d^9 lead to strong distortion.

The above three postulates (i), (ii) and (iii) which describe the effect of unsymmetrical t_{2g} or e_g orbitals on the shape of an octahedral complex is known as Jahn–Teller effect.

Explanation : The crystal field theory explains Jahn–Teller effect. This can be illustrated by con-sidering the distortion produced by the presence of e_g orbitals in a complex of Cu^{2+} ion (d^9 ion). This ion possesses the configuration $t_{2g}^{\ 6}\ e_g^{\ 3}$ in both the fields. Thus, there are two possible configurations L :

$$\text{I}\quad t_{2g}^{\ 6}\ (d_z^{\ 2})^2\ (dx^2 - y^2)^1$$

$$\text{II}\quad t_{2g}^{\ 6}\ (d_z^{\ 2})^2\ (dx^2 - y^2)^1$$

As both the configurations (t_{2g}) orbitals) are completely field, it means that asymmetry is due to the incomplete filling of e_g orbitals. Thus, distortion is mainly due to the repulsion of ligands by the electrons occupying e_g orbitals.

In configuration I the doubly occupied $d_z^{\ 2}$ orbital offers greater shielding to the nuclear charge of the copper than the half filled $dx^2 - y^2$ orbitals. The result of this is that :

(i) The ligands along z-axis which realise lower effective nuclear charge tend to move away from Cu^{2+} nucleus, and

(ii) The ligands along xy plane which experience a higher effective nuclear charge tend to move closer.

From the above two observations (i) and (ii), it is evident that the complex of copper should have four short and two long metal ligand bonds. If the electronic configuration II is accepted, it is expected that there will be a distortion but in the opposite direction in which the configuration I is predicting. According to conlguration II, the net result is :

(i) The ligands along z-axis tend to move closer to the nucleus and

(ii) The ligands along xy plane would move away.

From the above two results (i) and (ii), it follows that the complex of copper should have four long and two short metal-ligands bonds. Theoretically, there is no explanation for deciding which of the two possible distortions is more likely. How-ever, experimental results have revealed that distortions are more common in those octahedral complexes in which there are four short and two long bonds. For example,

(i) In the cupric chloride crystal each Cu^{2+} ion is surrounded by six Cl^- ions. Four of them are lying at a distance of 2.30 Å and the other two are at 2.95 Å away.

(ii) In cupric fluoride crystal F^- ions are lying at a distance of 1.93Å and the remaining two F^- ion are at 2.27 Å away.

We will now consider the effect of distortion due to Jahn–Teller effect on the energy of Cu^{2+} ion (d^9).

In splitting of the more stable octahedral distortion corresponding to configuration I $t_{2g}6$ $(dz)^2$ $(dx^2 - y^2)^1$ is being considered. δ_1 and δ_2 represent the splitting of the $e_g{}^{3-}$ and – levels respectively. Both δ_1 and δ_2 are much smaller as compared with Δ_0 and also, δ_2 is much smaller than δ_1 *i.e.*,

$$\Delta_0 >>> \delta_1 > \delta_2$$

Under these conditions both t_{2g} and e_g orbitals are split. For t_{2g} electrons there is no net energy exchange because four electrons (d_{yx} and d_{zx}) are stabilised by $4 \times (-1/3\ \delta_2) = -4/3\ \delta_2$ and the remaining two electrons (d_{zy}) get stabilised by $(2 \times \delta_2 + 2/3\ \delta_2) = 4/3\ \delta_2$. Thus,

Net energy for t_{2g} electrons = energy gain + energy loss

or Net energy gain for t_{2g} electrons $= -4/3\ \delta_2 + (4/3\ \delta_2) = 0$.

In the splitting at e_g levels, two electrons are stabilised by $2 \times \delta_1$ while one electron is destabilised by $1 \times = \delta_1$, *i.e.*,

Net energy gain for e_g electrons = energy gain + energy loss

$$= +1/2\ \delta_1 - \delta_1 = -\delta_1/2.$$

For the above simple calculations, the net energy change is $\delta_1/2$ which provides the driving force of distortion. This energy is called *John-Teller stabilisation energy.*

Failures of C.F.T.

(i) In CFT model metal-ligand bonding is taken to be purely electrostatic which is quite un-realistic. It provides no account of partial covalent character of the metal- ligand bonds.

(ii) The theory cannot explain for the relative strengths of ligands. For example, it provides no account why OH^- has a smaller splitting than H_2O.

(iii) CFT considers only the metal ion d-orbitals without giving any consideration to other metal orbitals such as s, p_x p_y and p_z and ligand π-orbitals.

(iv) CFT is unable to consider the possibility of the existence of π-bonding in complexes despite of its frequent existence in complexes of metals in low or high oxidation states.

(v) Crystal field theory keeps the total emphasis on the metal orbitals without consideration to the ligand orbitals. Therefore, all properties dependent upon the ligand orbitals and their interaction with metal orbitals are not at all explained.

Comparison of Valence Bond Theory and Crystal Field Theory

Some of the properties of complexes which could not be explained on the basis of valence bond theory are satisfactorily explained by crystal field theory. CFT is thus definitely an improvement over VBT. The following merits of CFT over VBT will prove that statement :

(i) CFT predicts a gradual change in magnetic properties of complexes rather than the abrupt change predicted by VBT.

(ii) In some complexes, when Δ_0 is very close to P, simple temperature changes may affect the magnetic properties of complexes. Thus, CFT provides theoretical basis for understanding and predicting the variation of magnetic moments with temperature as well as the detailed magnetic properties of complexes. This is just in contrast to VBT which cannot predict or explain magnetic behaviour beyond the level of specifying the number of unpaired electrons.

(iii) Though the assumptions inherent in VBT and CFT are vastly different, the main difference lies in their description of the orbitals not occupied in the low spin states. VBT forbids their use as they are involved in forming hybrid orbitals, while CFT strongly discourages their use as they are repelled by the ligands.

(iv) According to VBT, the bond between the metal and the ligand is covalent, while according to CFT it is purely ionic. The bond is now considered to have both ionic and covalent character. Unlike valence bond theory.

(v) CFT provides a framework for the ready interpretation of such phenomena as tetragonal distortions.

(vi) CFT provides satisfactory explanation for the colour of transition metal complexes, *i.e.*, spectral properties of complexes.

(vii) CFT can semiquantitatively explain certain thermodynamic and kinetic properties.

(viii) CFT makes possible a clear understanding of stereochemical properties of complexes.

6. Ligand Field Theory

Recently based on crystal field theory of **Bethe** (1929) and **Van Vleck** (1931–1935), a new approach known as ligand field theory was postulated.

In the crystal field theory, attraction between central metal and ligand in a complex is regarded as purely elctrostatic, *i.e.*, the bonding between the central metal ion and ligands is purely ionic. Thus, in crystal field theory covalent character of bond between metal and ligand is not taken into account. However, the enough evidence is available which suggests there is some measure of covalent bonding in complexes. The various experimental evidences are as follows.

(i) *Most direct evidence is obtained from electron* ***resonance spectrum*** *(ESR) of complexes. ESR spectrum of* $[IrCl_6]^{2-}$ *clearly shows hyperfine splitting indicating the delocalization of d electrons onto six chlorines. Such study of other complexes too gives similar results.*

(ii) ***Nuclear magnetic resonance (NMR)*** *studies of the complexes like* $KMnF_3$ *and* $KNiF_3$, *definitely indicate that the metal* t_{2g} *and* e_g *electrons pass a fraction of time around the fluorine nuclei.*

(iii) ***Nuclear quadrupole resonance (NQR)*** *spectra of square planar halide complexes of divalent platinum and palladium suggest considerable covalency in the metal-ligand bonds.*

(iv) ***The unusually large absorption band intensities*** *observed for tetrahedral complexes like* $[CoCl_4]^{2-}$ *have been interpreted as due to appreciable covalent nature of the metal-ligand bonds.*

The ligand field theory is eventually the same as pure crystal field theory but covalent character being taken into account.

When the orbitals overlap, *i.e.*, covalent character is excessive as in metal complexes of carbon monoxide or the isocyanide, the *molecular orbital theory* gives a more complete explanation of the *metal-ligand bonding*.

Ligand field theory differs from the valence bond theory by recognising the existence of antibonding as well as the bonding and non-bonding orbitals. For the present, the ligand field theory seems to offer the most practical approach to bonding in co-ordination compounds.

In ligand field theory, it is assumed that molecular orbitals are formed by the overlap of orbitals from the ligand with the atomic orbitals of the central atom. (The procedure is similar as adopted in case of simple molecules). The molecular orbitals thus formed may be of a *bonding, antibonding or a non-bonding character.*

The antibonding orbitals are similar to the bonding orbitals except that the antibonding orbitals lie higher in energy and have nodes or regions of low electron density between the central atom and the ligands. The antibonding orbitals are of interest because it is these into which electrons may be excited from the *tis* orbitals by absorptions of energy. The non-bonding orbitals in octahedral complexes are simply the *dxy, dxz* and *dyz* orbitals of the central atom.

As the number of molecular orbitals formed is always equal to the number of atomic orbitals taking part in the overlappings and this number is quite large for complexation processes. The MO energy level diagrams for metal complexes are highly complicated. The fact that σ as well as π (both types of overlaps are involved in the coordination process) add more complication to these diagrams.

The three common types of complexes are octahedral, tetrahedral and square planar. But we shall discuss octahedral complexes.

OCTAHEDRAL COMPLEXES

(a) Central Metal Ion Orbitals. In an octahedral complex the metal ion is surrounded by six ligands placed along the $+x$, $-x$, $+y$, $-y$, $+z$ and $-z$ axes. We know that there are 9 atomic orbitals on the central metal atom if it belongs to $3d$ series. These are one outer 4s, three outer p ($4p_x$, $4p_y$ and $4p_z$) and five inner $3d$ ($3d_x^2 - y^2$, $3d_z^2$ $3d_{xy}$, $3d_{yz}$, $3d_{zx}$) atomic orbitals. These orbitals can be divided into two groups.

(i) The first group includes 4s, $3p_x$, $4p_y$, $4p_z$ and $3d_x - y^2$ atomic orbitals. These orbitals have their lobes lying along the axes. The orbitals by overlapping with ligand orbitals form six metal-ligand σ bonds or molecular orbitals.

(ii) The remaining three atomic orbitals of central metal atom do not participate in σ-bonding and hence remain non-bonding. These can, however, take part in π-bonding in octahedral complexes by overlapping sidewise with π-orbitals of the same six ligands that are approach-ing the central metal ion from both sides along + x, – x, + y, – y + z and – z axes.

On the basis of the group theory, the metal orbitals are usually classified as

$dx^2 - y^2$, dz^2	e_g	or	Eg
s	a_{1g}	or	A_{1g}
p_x, p_y, and p_z	t_{1u}	or	T_{1u}
d_{xy}, d_{yz} and d_{zx}	t_{2g}	or	T_{2g}

In the above table, it is to be remembered that.

(i) a or A orbitals are always non–degenerate.

(ii) e or E orbitals, are double degenerate.

(iii) t or T orbitals are triple degenerate.

(iv) g expresses the gerade nature of orbitals and

(v) u expresses the ungerade nature of orbitals.

An orbital is called a gerade orbital if on traversing equal distances in a straight line in opposite directions from the centre of symmetry, the sign of wave function does not change. Examples of such

(b) Formation of Group or Composite Ligand Orbitals : It is clear that each of the metal orbitals would be in a position to overlap simultaneously with several ligand orbitals which may in turn belong to different ligands. But the same is not true for ligand orbitals. Before overlapping, the six ligand orbitals combine together linearly to form six group σ-orbitals that would be capable of overlapping with the central metal ion.

The determination of such linear combination of ligand o-orbitals is made by inspection method which is a straight for-ward and commonsense method. The linear combinations of ligand or-bitals are called composite ligand orbitals. The positions of six ligand σ–orbitals in an octahedral complex. The formation of group ligand σ–orbitals is shown pictorially.

(i) If the +ve lobes of a orbitals of all the six ligands in an octahedral complex are pointing towards the central metal ion, then the composite orbital, signified by $(\sigma_x + \sigma_{-x} + \sigma_y + \sigma_{-y} + \sigma_{-z})$ is capable of overlapping with the s orbitals of the metal cation. This composite orbital is designated as Σ_a and its value when normalised is as follows :

$$\Sigma_a = (\sigma_x + \sigma_{-x} + \sigma_y - \sigma_{-y} +. \sigma_z + \sigma_{z})$$

(ii) Similar to above, for overlapping with p orbitals, the composite orbitals can be find out. For px, it would be $\sigma_x - \sigma_{-x}$. Similarly for p_y and p_z these would be $\sigma_y - \sigma_{-y}$ and $\sigma_z - \sigma_{-z}$. The composite orbitals p_x, p_y and p_z are designated as Σ_x, Σ_y, and Σ_z, respectively. Their values, when normalised, are follows.

$$\Sigma_x = (\sigma_x - \sigma_{-x})$$

$$\Sigma_y = (\sigma_y - \sigma_{-y})$$

$$\Sigma_z = (\sigma_z - \sigma_{-z})$$

(iii) For $dx^2 - y^2$, the composite orbital would be $(\sigma_x - \sigma_{-y} - \sigma_y - \sigma_{-y})$. Its normalised value is

$$\Sigma_x{}^2 - y^2 = (\sigma_x + \sigma_{-x} - \sigma_y - \sigma_{-y})$$

(iv) For $d_z{}^2$, the composite orbitals would be $(2s_z + 2s_{-z} - r_x - s_{-x}\ s_y - s_{-y})$ Its normalised value is

$$\Sigma_z{}^2 = (2\sigma_z + 2\sigma_{-z} \quad \sigma_x - \sigma_{-x} - \sigma_x - \sigma_y - \sigma_{-y})$$

In the above equation the value of 2 before σ_z and σ_{-z} is the value of C $(C\psi_M = C_1\psi_1 + C_2\psi_2)$.

The value of C decides the magnitude of contribution of that atomic orbital towards the formation of molecular orbital.

(c) Formation of Molecular Orbitals. In the next step, six atomic orbitals of the central metal ion.

$4s$, $4p_x$, $4p_y$, $4p_z$, $3d_z^2$ and $3d_x^2$ overlap with six groups of composite ligand orbitals formed as in step (b)) forming six sigma bonding and six sigma antibonding molecular orbitals. The bonding molecular orbitals are designated as σ^b – MO's whereas the antibonding molecular orbitals are designated as σ^* – MO's. Thus,

(i) 4_S *and* Σ_a *on overlapping give and MO's.*

(ii) $4px$ *and* Σ_x *on overlapping give and MO's.*

(iii) $4py$ *and* Σ_y *on overlapping give and MO's,*

(iv) $4p_z$ *and* Σ_z *on overlapping give and MO's.*

(v) $3dx^2 - y^2$ *and* $\Sigma x^2 - y^2$ *on overlapping give* $\sigma^b\ x^2 - y^2$ *and* $\sigma^*\ x^2 - y^2$ *MO's.*

(vi) $3d^2$, *and on overlapping 'give* $\sigma^b z^2$ *and* $\sigma^* z^2$ *molecular orbitals.*

The remaining three orbitals on central metal ion, *i.e.*, $3d_{xy}$, $3d_{yz}$ and $3d_{zx}$ are non-bonding orbitals. Thus, there are is molecular orbitals available for electron fill-ing. All the fifteen molecular orbitals. The number of electrons which each group of or-bitals can hold is given in brackets.

(d) Filling of Molecular orbitals : Filling of the molecular orbitals occurs according to the Aufbau's principle. With d^n complexes, the t_{2g} and $e_g{}^*$ orbitals will fill up in just the same way as in crystal field theory. Let us illustrate this filling by considering a d^3 octahedral complex like $[Cr(NH_3J)_6]^{3+}$. In this complex, there is a total of 15 electrons (twelve from six metal ligand orbitals overlapping and three from metal *d* orbitals). These electrons are to be accommodated. Out of 15 electrons, twelve electrons are accommodated in the six bonding MO, *i.e.*, A_{1g} T_{1g} and E_g whereas the remaining three go to the triply degenerate *i.e.*, non-bonding T_{2g} orbitals in the unpaired state. The energy difference between T_{2g} and degenerate sets of orbitals is referred to as Δ.

This is the same as Δ_0 of the complexes described in crystal field theory. The only difference is that in ligand field theory, the value of Δ arises due to the different amount of electrostatic repulsion suffered by the various metal *d* electrons from the ligands whereas in molecular orbitals theory, Δ arises due to the metal-ligand covalent bonding which produces one set of non-bonding orbital T_{2g} and another of high energy antibonding orbitals () Thus, in molecular orbital theory, the concept of high and low spin complexes remains same as in crystal field theory. The magnitude of Δ decides the choice between the high and low-spin complexes. For example, in the hexafluoro complex of Co^{3+} the value of Δ is small and the distribution of electrons between and T_{2g} and occurs as , . This makes it a high spin complex.

This distribution also explains why the Co–F bonds in the complex are not very strong. The reason for this is that the presence of two electrons in the an-tibonding orbitals reduces the strength of Co–F bonds. It is also important to remember that high spin complexes contain electrons in the antibonding orbitals This also explains why such complexes are relatively less stable.

(e) π-Bonding in Octahedral Complexes : The metal orbitals used for π bonding are as follows

$$d_{xy}, d_{yz}, d_{zx} - t_{2g} \text{ or } T_{2g}$$

and

$$p_x, p_y, p_z - t_{1u} \text{ or } T_{1u}$$

From the above it follows that t_{1u} orbitals are involved in o as well as π-overlaps whereas t_{2g} involve in π-overlaps. The ligand orbitals which take part in the π-overlaps are generally *p*π or *d*π orbitals. In the positions of various *p*π orbitals of the ligands are shown. In these the arrow indicates the direction of positive lobes of these orbitals. Each ligand will have two such *p*π orbitals at right angles to each other.

The composite ligand orbitals for these can be evolved as before. For example, the composite ligand orbitals of the *px (t1u)* orbitals of the metal would be $\pi_{3x} + \pi_{4x} + \pi_{5x} + \pi_{6x}$. For the d_{xz} t_{2g} orbital of the metal, the composite ligand orbital would be $(\pi_{1z} + \pi_{2z} + \pi_{5z} + \pi_{6x})$. Such types of π-orbitals are present in the ligand like oxides and fluorides. These orbitals have lower energy than the metal-π orbitals. In π-overlaps, eighteen molecular orbitals (six metals orbitals and twelve ligand orbitals, 2π-orbitals on each ligand) are formed. Out of these six are bonding and $[T_{2g}$ (3) and $T_1\pi$ (3) and six are non-bonding $[T_{2g}\pi^*$ (3) and $T_{2g}\pi$ (3)]. It is important to remember that T_{1g}^* and T_{1g}^* have the same ligand combination as T_{1g} and $T_{2g}\pi$, but with every other sign in the linear combination reversed. Similarly, $T_{1g}\pi^*$ and $T_{1u}\pi^*$ have the same combination as $T_{2g}\pi$ and T_{1u} but with every sign reversed.

The magnitude of Δ increases according to the nature of the ligands in the following order Strong π donor < weak π–donor < negligible

π-interaction < weak π–acceptor < strong π-acceptor.

This confirms the order of spectrochemical series.

SPECTROCHEMICAL SERIES

One of the factors which affects the degree of splitting of d–orbitals (10D or Δ) is the *nature of the ligands.* The value of 10Dq can be equated to the energy observed as the maximum in the absorption spectra of complexes as explained earlier for $[Ti(H_2O)_6]^{3+}$. ON this basis, ligands may be arranged in order of their field strength, *i.e.*, the order of energy of the first absorption bands of the complex.

$I^- < Br^- < S^{2-} < SCN^- < Cr^- < \; < F^- < OH^- < Ox^{2-} <$

$H_2O < NCS^- < NH_3 <$ en < dipy < phen < $< CN^- <$ CO

This arrangement of ligands in the order of *increasing field strength* is called Fajans-Tschida *Spectrochemical series* because it is derived from spectral studies of the complexes. The ligands on the right side of the series provide the strongest field (high 10Dq) are the most effective in forcing the cations into low spin states whereas the ligands on the left, provide weak field (low 10Dq) add usually give high spin complexes.

For example we may consider the formation of two oc-tahedral complexes of Fe^{3+} $[Fe(H_2O)_6]^{3+}$ and $[Fe(CN)_6]^{3-}$ where Fe^{3+} has d^5 configuration. Since H_2O is a weaker ligand than CN^- ion, the value of splitting energy (10 Dq) is smaller in $[Fe(H_2O)_6]^{3+}$ than in $[Fe(CN)_6]$. Hence, the five d electrons of Fe^{3+} will occupy all the t_{2g}, and e_g levels without pairing, resulting in a high spin complex with 5 unpaired electrons. Thus $[Fe(CN)_6]^{3-}$ is a highly paramagnetic complex. In case of $[Fe(CN)_6]^{3-}$ the value of 10 Dq is sufficiently high to force pairing of t_{2g} electrons. Thus in $[Fe(CN)_6]^{3-}$ there are two pairs of electrons in two t_{2g} orbitals and single electron in one t_{2g} orbital. Hence it is also paramagnetic in nature but it is less paramagnetic as compared to $[Fe(H_2O)_6]^{3+}$. The effect of the order of increasing ligand field strength in spectrochemical series, will be clear from the absorption spectra of some complexes of Co^{3+} and Cr^{3+}.

We can see that there is a steady, progress in frequency of the absorption maxima as the ligands change with the linking atoms progressing Cl $\rightarrow$ S $\rightarrow$ O $\rightarrow$ N $\rightarrow$ C corresponding to a progres-sive increase in the value of 10 Dq.

Though spectrochemical series allows us to rationalise differences in spectra, it creates many serious problems which could not be accounted for, on the basis of crystal field theory. If the ligands arc considered as point charges or dipoles (as as-sumed by CFT) then anionic ligands should bring out more crystai field splitting than the neutral molecules. According to CFT, among negative ions F^- is expected to be a stronger ligand than CN^- and among neutral molecules H_2O is expected to be stronger than NH_3 (H_2O has higher dipole mo-ment than NH_3). But in the spectrochemical series, their positions are just the reverse. The position of these ligands can be explained only when π bond-ing is assumed between the central metal and the ligand (*Molecular orbital theory*).

Since CFT com-pletely neglects π bonding, it cannot account for the position of all the ligands in the spectrochemical series. This failure to interpret the ligand field strength remains as one of the major drawbacks of the crystal field theory.

CO-ORDINATION COMPOUNDS

Profound changes occur in the properties of a metal ion on complex formation. These changes are generally reflected in solubility, colour, stability, magnetic properties, etc. These changes in properties find manifold applications.

1. Dyes : The characteristic colours of co-ordination compounds would indicate their uses as dyes. Alfred Werner demonstrated the relation of co-ordination with dyeing and he showed that several com-pounds which were capable of forming metal-chelate compounds were able to dye cloth pretreated with ferric hydroxide.

The first complete study of co-ordination compounds as dyes was given by G.T. Morgan and his co-workers in the early 1920.

2. As catalysts : Many enzymes, which serve as the catalysts in living system, are coordination compounds. For example, the decomposition of hydrogen peroxide is catalysed by many things, including iron compounds.

$$2H_2O_2 \quad 2II_2O + O_2.$$

This is evident from the following points

(i) *Ordinary hydrated ferric ion has a relative activity of unity,*

(ii) *A co-ordination compound of iron, the he me (human blood), has a relative activity of one thousand, and*

(iii) *Catalase, a heme surrounded by a complicated protein structure, has a relative catalytic activity of ten billion.*

3. Analytical Applications : The basis of qualitative detection of metal ions is the complexation reactions of such ions. Common examples of qualitative detection of metal ions by way of complexation reactions include detection of nickel (II), as red coloured bis (dimethylglyoximato) nickel (II), of cobalt as blue coloured tetrakis (isothiocyanato) cobalt

(II), $[Co(NCS)_4]^{2-}$, of iron as blood red hexakis (isothiocyan-to) iron (III), $[Fe(NCS)]_3^-$ etc. These are as follows.

(a) Aluminium and zinc salt solutions form with excess of NaOH, soluble sodium aluminate and sodium zincate.

$$AlCl_3 + 3NaOH \quad + 3NaCl$$

$$Al(OH)_3 + NaOH \quad + 2H_2O$$

$$ZnCl_2 + 2NaOH \quad + 2NaCl$$

$$Zn(OH)_2 + 2NaOH \quad + 2H_2O$$

Similarly, chromium hydroxide dissolves in excess of NaOH forming sodium chromite.

$$CrCl_3 + 3NaOH \quad Cr(OH)_3 + 3NaCl$$

$$2Cr(OH)_3 + 2NaOH \quad + 4H_2O$$

Chromium salts in the presence of oxidising agents such as Br_2 water, H_2O_2, form a soluble yellow chromate ion.

$$2CrCl_3 + 10NaOH + O \quad + 6NaCl + 5H2O$$

These reactions have been utilised for the separation of aluminium from iron and chromium and zinc from manganese.

(b) Iodine is having feeble solubility in water. However, its solution is prepared by dissolving iodine in a solution of potassium iodide. This solubility is due to the formation of complex KI_3.

$$KI + I_2 \quad KI_3 \quad K^+ +$$

When potassium iodide solution is added to mercuric chloride solution, a scarlet red ppt. of mercuric iodide is obtained which dissolves in excess of KI forming K_2HgI_4.

$$HgCl_2 + 2KI \quad HgI_2 + 2KCl$$

$$HgI_2 + 2KI$$

When this complex is added to sodium hydroxide, the solution is known as Nessler's reagent which is employed for the *detection of ammonia.* It forms brown colour or precipitate with ammonia.

$$2K_2HgI_4 + 3NaOH + NH_4OH \rightarrow \begin{matrix} NH_3 \\ Hg \\ O \\ Hg \\ I \end{matrix} + 4KI + 3NaI + 3H_2O$$

Iodine of Millon's base

(c) When potassium cyanide is added to copper and cadmium salts solution, both copper and cadmium from complexes. These two complexes have different stabilities.

$$CuSO_4 + 2KCN \quad + K_2SO_4$$

$$2Cu(CN)_2 + 6KCN \quad + (CN)_2$$

$$CdSO_4 + 3KCN \quad Cd(CN)_2 + K_2SO_4$$

$$Cd(CN)_2 + 2KCN$$

When hydrogen sulphide is passed in these two solutions of complexes, cadmium is only precipitated because the copper complex is much more stable than the cadmium complex.

$$K_3Cu(CN)_4 \quad 3K+ + \ [Cu(CN)_4]^{3-}$$

$$[Cu^+ + (CN)_4]^{3-} \quad Cu^+ + 4CN^- \text{ (Negligible dissociation)}$$

$$K_2[Cd(CN)_4] \quad 2K^+ + [Cd(CN)_4]^{2-}$$

$$[Cd(CN)_4]^{2-} \quad Cd^{2+} + 4CN^- \text{ (High dissociation)}$$

(d) Separation of nickel and cobalt : When potassium cyanide is added to a mixture of cobalt and nickel salts, potassium cobalto cyanide is formed whereas nickel only forms nickel cyanide. Potassium cobalto cyanide with bromine and alkali is converted into potassium cobalticyanide whereas nickel cyanide with bromine and NaOH yields a black ppt. of nickel oxide.

$$CoCl_2 + 2KCN \quad Co(CN)_2 + 2KCl$$

$$Co(CN)_2 + 4KCN \quad K_4[Co(CN)_6]$$

$$2K_4[Co(CN)_6] + H_2O \mid O \qquad 2K_3[Co(CN)_6] + 2KOH$$

$$NiCl_2 + 2KCN \quad Ni(CN)_2 + 2KCl$$

(e) Reaction with dimethyl glyoxime : Ni^{2+} form a red ppt. with dimethyl glyoxime. This reaction has been used for the separation of nickel from cobalt in the group IV.

$$+ 2NH_4OH + NiCl_2$$

(f) Reactions with yellow ammonium sulphide : The reaction of yellow ammonium sulphide with sulphides of IInd group cations is used for the separation of IIA from IIB. IIB cations form soluble complexes while of IIA cations do not form.

Arsenic,

$$As_2S_3 + 3(NH_4)_2S$$

$As_2S_5 + 3(NH_4)_2S$

Antimony,

$Sb_2S_3 + (3NH_4)_2S$

$Sb_2S_5 + 3(NH_4)_2S$

Tin

$SnS + (NH_4)_2S$

$SnS_2 + (NH_4)_2S$

These thioarsenite, thioarsenate, thioantimonite, thioantimonite, thiostannanite and thiostannate are soluble complexes from which original sulphides of arsenic, antimony and tin can be ppted by making the solution acidic with dilute hydrochloric acid.

(f) Complexes are usually formed to keep a metal ion in solution or to dissolve precipitate and bring the metal ion back into solution. For example, AgCl dissolves on addition of ammonia solution and forms the complex ion, diamine silver (I) ion.

$AgCl + 2NH_3 \quad [Ag(NH_3)_2]^+ + Cl^-$

(g) Formation of complexes with characteristically coloured solutions or precipitates is used for the detection of metal ion in solution. For example, Cu^{2+} ion can be detected in solution by the dark blue coloured solution it forms on addition of ammonia.

$Cu^{2+} + 4NH_3$

(h) Use in gravimetric determination : Inner complexes are often insoluble in aqueous medium but soluble in organic solvents. The formation of such chelates often needs suitable pH range and many metal ions can be quantitatively precipitated and metal ions determined gravimetrically. Some applications include estimation of aluminium as yellow coloured tris (8–hydroxy quinolonate), aluminium (III) and copper (II) as light green coloured bis-(quinolonate) aluminium (III) and copper (II). Some complexes are unstable at the drying temperature (120–150°C) and such complexes are ignited to metal oxides.

(i) Use of organic sequestering agents in removal of interference in gravimetric estimations :

An organic ligand may read with more than one metal ion to form sparingly soluble precipitates. In such cases direct estimation of a particular metal ion in the presence of interfering ions is not possible. A strong sequestering agent like EDTA is of great importance in making interfering ions ineffective, and thus making the precipitating ligand almost specific for a particular metal ion. For example, in the presence of EDTA, beryllium may be precipitated with ammonia in presence of chromium, cobalt, cadmium, iron, copper, lead, manganese, zinc, aluminium, bismuth etc.

(j) Solvent extraction and separation of metal ions through complexation : Both iron (III) and aluminium (III) form complexes with 8-hydroxy quinoline. Iron (III) complex is easily soluble in chloroform. Iron (III) can be separated from aluminium (III) around pH 3.0 by the addition of 8–hydroxy–quinoline (oxine), followed by extraction with chloroform.

(k) Complexometric titrations : The sequestering effect of EDTA has led to the development of some versatile titrimetric reagents for the estimation of metal ions. Mostly EDTA titrations are performed with indicators, which are chelating agents whose metal complexes have different colours from the reagent.

The indicators most commonly employed are Erichrome black T, murexide, etc. Such complexometric titrations are used in the determination of water hardness. This procedure affords various advantages as compared to old and laborious palmitate titrations.

(I) Spectrophotometric estimation of micro amounts of metal ions. A large number of metal complexes are intensely coloured which provide a basis for their spectrophotometric determination when present in trace amounts. For example, iron (II) forms dark red complex with o-phenanthroline over a wide range of pH 2–9. Trace amounts ($\sim 10^{-6}$ gm/ml.) of iron (II) can be estimated by measuring the optical density of the complex at 5.5 mμ.

3. Photography : Formation of $[Ag(S_2O_3)_2]^{3-}$ complex ion has an important role in photography. After the exposed film is developed, sodium thiosulphate solution is used to dissolve the unreduced silver halide.

$$AgBr + 2S_2O_3^{2-} \quad [Ag(S_2O_3)_2]^{3-} + Br^-$$

4. Extraction of Metals from Ores : The noble metals like Ag and Au are extratced from their ores through the formation of cyanide complexes like $[Ag(CN)_2]^-$ and $[AuCl_4]^-$ respectively.

5. Artificial Silk : Schweitzer's reagent, *i.e.*, tetrammine copper (II) hydroxide $[Cu(NH_3)_4](OH)_2$ is used as a solvent for cellulose during the manufacturing of artificial silk.

6. Biological Applications : A number of metal complexes are of biological importance. In particular (i) haemoglobin in the red blood cells contains an iron-porphyrin complex, (ii) chlorophyll in green plants contains magnesium-porphyrin complex, (iii) vitamin B^{12} is a cobalt complex. Body contains a number of compounds like adrenaline, citric acid and cortisone which form unwanted complexes with metals like lead, copper etc. and prevent normal metabolism . Lead poisoning and copper poisoning are treated by injecting EDTA, so that metal–EDTA complex is excreted in the urine.

SOME SOLVED PROBLEMS

Problem 1:

A mixture is to be assayed for penicillin. You add 10.4 *mg of penicillin of specific activity* 0.405 $\mu C\ mg^{-1}$. *For this mixture you are able to isolate only* 0.3 *mg of pure crystalline penicillin, and you determine its specific activity to be* 0.035 $mC\ mg^{-1}$. *What was the penicillin content of the original sample ?*

Solution:

From the given data,

$$m' = 10.0 \text{ mg}$$

$$S' = 0.405\ \mu C\ mg^{-1}$$

$$S = 0.035\ \mu\ C\ mg^{-1}$$

Substituting these values in following eq., we get

$$m = m'\left[\frac{S'}{S} - 1\right] \text{ mg.} \qquad ...(1)$$

$$= 10.0\left[\frac{0.405}{0.035} - 1\right] \text{ mg} = 105.71 \text{ mg } \textbf{Ans.}$$

Problem 2:

A *five kilogram batch of crude penicillin is assayed by isotope dilution. To one gm of sample of the batch was added 10 mg of pure penicillin having an activity of 10,600 c.p.m.; only 1.40 mg of pure penicillin was recovered having an activity of 280 c.p.m. What is the penicillin content of the batch ?*

***Solution*:**

$$m' = 10.0 \text{ mg}$$

$$S' = \frac{10600}{10} \text{ c.p.m. mg}^{-1}$$

and $$S = \frac{280}{1.40} \text{ c.p.m. mg}^{-1}$$

From equation (1), it follows that

$$m = 10.0 \left[\frac{\frac{10600}{10}}{\frac{280}{1.40}} - 1 \right] \text{mg}$$

$$= 10.0 \left(\frac{1060}{200} - 1 \right) \text{mg}$$

$$= 10.0 \left(\frac{1060 - 200}{200} \right) \text{mg}$$

$$= \frac{10.0 \times 860}{200} \text{ mg} = 43 \text{ mg}$$ **Ans.**

One gm of the batch contains 43 mg of penicillin, 5000 gm (5 kg.) of the batch contain 43 × 5000 mg of penicillin or 5 kg of the batch contain 21,5000 mg or 215 gm of penicillin

Problem 3:

A piece of wood recovered in an excavation has 25.6% as much C^{14} as ordinary wood today. When did this piece get buried ?

***Solution* :**

C_0= 100% : Ct = 25.6%; $t_{1/2}$ = 5760 years.

Since $$K = \frac{0.693}{t_{1/2}} = 1.203 \times 10^{-4} \text{ years}$$

$$Kt = \ln \frac{C_0}{C_t}$$

$$t = \frac{2.303}{1.203 \times 10^{-4}} \log \frac{100}{25.6} = 11330 \text{ years } \textbf{Ans.}$$

Problem 4:

A sample of carbon derived from one of the dead sea scrolls is found to be decaying at the rate of 11.5 *disintegration per min. per gram of carbon. Estimate the age of dead sea scrolls.* ($T_{1/2}$ for C^{14} = 5568 *yrs, carbon from living plants disintegrates at the rate of 15.3 disintegrations per min per gram*).

Solution: We know that,

$$\lambda = \frac{0.693}{T_{1/2}} = \frac{0.693}{5568} = 1.244 \times 10^{-4} \text{ yr}^{-r}$$

$$t = \frac{2.303}{1.244 \times 10^{-4}} \log \frac{15.3}{11.5}$$

$$= 2.2956 \times 10^{3} \text{ yrs}^{-1}. \textbf{ Ans.}$$

Problem 5:

The ratio by weight of $^{206}_{82}Pb$ *and* $^{238}_{92}U$ *in a uranium mineral is* 0.2. *If* $T_{1/2}$ *for* $^{238}_{92}U$ *is* 4.5 × 10⁹ *yrs, calculate the age of the mineral.*

Solution:

The ratio by weight of $^{206}_{82}Pb$ and $^{238}_{92}U$ = 0.2 : 1

∴ The ratio by moles of $^{206}_{82}Pb$ and $^{238}_{92}U$ = $\frac{0.2}{206} : \frac{1}{238}$

∴ 0.23 : 1

It means originally the mineral contains

$$= 1 + 0.231 = 1.231 \text{ g atom of } ^{238}_{92}U$$

We known that $k = \frac{0.693}{T_{1/2}} = \frac{0.693}{4.5 \times 10^{9} \text{ yrs}}$

$$= 1.54 \times 10^{-10} \text{ yr}^{-1}$$

$$\therefore t = \frac{2.303}{k} \log \frac{N_0}{N} = \frac{2.303}{1.54 \times 10^{-10}} \log \frac{1.231}{1}$$

$$= 1.58 \times 10^{9} \text{ yrs } \textbf{Ans.}$$

Problem 6:

To a protein hydrolysate was added 10.1 *mg of labelled alanine of specific activity 128 c.m.p. mg*$^{-1}$ *measured in a particular counting arrangement. A sample of pure alanine isolated from the mixture was found to have a specific activity of 68.3 c.p.m. mg*$^{-1}$ *when measured in the same counting arrangement. What was weight of alanine present in the hydrolysate ?*

Solution:

From the given data.

$$m' = 10.1 \text{mg}$$

$$S' = 128 \text{ c.p.m. mg}^{-1}$$

$$S = 68.3 \text{ c.p.m. mg}^{-1}$$

On substitution of these value in the equation, we get

$$m = m'\left[\frac{S'}{S} - 1\right] \text{ mg}$$

$$= 10.1\left[\frac{128}{68.3} - 1\right] = 8.83 \text{ mg.}$$

The weight of alanine in the hydrolysate

= 8.83 mg. **Ans.**